The Relevance of Financial Literacy

Edoardo Lanciano

The Relevance of Financial Literacy

Theory, Research, and Metrics

palgrave
macmillan

Edoardo Lanciano
University Roma Tre
Rome, Italy

ISBN 978-3-031-93396-7 ISBN 978-3-031-93397-4 (eBook)
https://doi.org/10.1007/978-3-031-93397-4

This Palgrave Macmillan imprint is published by the registered company Springer Nature Switzerland AG
The registered company address is: Gewerbestrasse 11, 6330 Cham, Switzerland

To my family
To Giulia

Foreword

The topics of financial literacy and related financial education initiatives have long been the focus of attention by many social actors, including international organizations, governments, the world of primary and secondary education, universities, academic research, and financial intermediaries. The growing complexity of the financial markets, the variety of financial services offered, and the crisis of public welfare systems make it necessary to intervene in formal education systems at all levels to enable the exercise of economic citizenship appropriate to these situations. In other words, public and private investments appear increasingly relevant and useful for improving the ability of individuals, communities, and businesses, especially smaller family businesses, to make various types of financial decisions. At an international level, the world of international organizations, banking and financial supervisory authorities, governments, education, research, and finance have published various contributions, in the form of articles, reports, and books on the topics of financial literacy and financial education policies and interventions. The perspectives, methodologies, and contents of the contributions are varied, and are aimed at different audiences, from professionals to citizens looking for good information about investment decisions, debt, choice of means of payment, choice of instruments to cover pure and financial risks, or investments for pension purposes.

We can define the literature on financial literacy and financial education policies and tools as a literature on good finance, as it aims to improve the knowledge of people, communities, and businesses that wish to improve their level of financial competence. Edoardo Lanciano's book is part of this strand of literature oriented towards good finance, completing and developing a doctoral course followed over the last 3 years, laying the foundations for further developments both in the field of applied research and in the equally important field of dissemination. This dual approach is present throughout the chapters of the book, making it useful both for those who work professionally in the field of measuring and improving levels of financial literacy and for those who want to learn about the topics outlined above in a rigorous and comprehensible way. In the final part, still following this double track, the topics of digitalization and environmental and social sustainability are addressed, which cross the social and economic life of the current world and the near future. A world where improving levels of financial literacy, through adequate and targeted financial education interventions aimed at less competent individuals and companies, can lead to the development of a more aware and effective economic citizenship.

March 2025

Daniele Previati
Professor of Financial Markets
and Institutions
Department of Business Studies
University Roma Tre
Rome, Italy

Acknowledgements I would like to sincerely express my deep gratitude to those who mainly shared with me, in various ways and with different roles, these years of study, research, and professional growth: Daniele Previati, Ornella Ricci, Franco Fiordelisi, Saverio Stentella Lopes, and Gianluca Santilli. Thank you for your support, guidance, and friendship.

Competing Interests The author has no competing interests to declare that are relevant to the content of this manuscript.

Introduction

This book provides a general overview on the topic of financial literacy, covering various theoretical and technical aspects such as definitions, global data, information on the level of financial literacy in society, metrics, critical perspectives, and analyses of the relationships with financial outcomes. Finally, recent research evidence on the role of financial literacy for modern challenges such as digitalization and sustainability are provided. This project aims to develop knowledge and awareness on the topic of financial literacy, by providing international evidence, research aspects and showing the effects on financial and social outcomes.

Financial literacy is indeed an issue of growing economic and social relevance, especially in recent times with the increasing need to be able to manage money autonomously, considering the progressive shift of financial system and financial services from the physical to the digital channel. Financial literacy is thus becoming essentially important for individual and social empowerment and well-being.

The first chapter aims to introduce the topic of financial literacy by providing and explaining the main definitions and analyzing the meaning and the different components of financial literacy. The chapter then presents general evidence of the level of financial literacy all around the world, available from international survey data such as OECD[1] and World Bank surveys, and descriptive analyses of the distribution of financial

[1] Organization of Economic Cooperation and Development.

literacy among different sociodemographic categories, highlighting gaps related to gender, education, or geographical origin, for example. Many people indeed lack of basic knowledge and understanding of essential financial concepts, and for this reason, financial literacy has become a long-term policy priority in several countries and economies, involving different actors. Finally, the chapter provides an overview of these actors with a relevant role for the general development of financial literacy in society, such as governments, supervisory authorities, financial intermediaries, educational institutions, and families, analyzing their contribution to the spread of financial education in society.

The second chapter is focused on technical and methodological aspects related to financial literacy, in particular explaining and comparing the different ways used in research to measure financial literacy on various targets, available from different survey experiences, from the Big Three questions to the OECD/INFE[2] toolkits. In addition, the chapter analyzes the main endogeneity issues arising in financial literacy studies, such as reverse causality and measurement error, and the main methods used to address them, such as instrumental variable approaches. In the following section, the chapter provides critical views on financial literacy, by discussing limitations of traditional research approaches and financial literacy indicators. Finally, the chapter discusses other possible research approaches including experimental techniques and alternative perspectives. In fact, the book focuses mainly on a mainstream approach to financial literacy based on specific measurements and analyzing techniques, while there are also other aspects related to different disciplines such as sociology, psychology, or pedagogy that have an impact on this topic.

The third chapter of the book analyzes the relationship between financial literacy and various specific financial outcomes related to individuals and entrepreneurs, such as financing needs, investment decisions, and risk protection strategies. The aim of this chapter is to highlight the relevance of financial literacy by reporting the main research evidence from distinguished studies conducted in several contexts, which show the effects of financial literacy on financial outcomes. This analysis is performed through a review of the most relevant studies available in the literature from the most expert scholars on this topic, providing an international

[2] International Network of Financial Education.

overview including studies from western and non-western contexts, which analyze these relationships.

The final chapter reports recent research evidence on the effect of financial literacy on the use of financial services delivered in digital form, and on the orientation towards sustainable finance decisions, which represent challenges with increasing relevance in society and in the financial services industry. In particular, after providing an overview of the available studies in the literature which are focused on these matters, and discussing the main results, this recent research evidence is focused on Italy, and thus performed using Italian data, provided by Bank of Italy and by the Italian National Financial Education Committee. This chapter provides evidence on the positive effect of individual levels of financial literacy on the adoption of fintech services and on sustainable investment decisions.

In addition to providing a general overview of the topic of financial literacy from a theoretical and empirical point of view, this project offers various potential contributions to the existing literature and thus aims to become a relevant reference for experts in the field, researchers, and students. The book provides an international overview about financial literacy and on research on financial literacy. First, it analyzes financial literacy levels worldwide, the relationship between financial literacy and various specific financial needs and outcomes, including evidence from studies conducted in different contexts around the world. Additionally, it provides global insights into main trends of financial literacy and financial education contributions, and compares different metrics and research approaches for measuring financial literacy, also providing a critical perspective. Finally, it focuses on contemporary challenges such as the relationship between financial literacy, digitalization, and sustainable finance.

Contents

1 Financial Literacy and Financial Education: An Overview 1
1.1 Introduction 2
1.2 Definitions 3
1.3 Evidence from Around the World: Survey Experiences 5
1.4 Sociodemographic Differences 11
1.5 Key Players 15
1.6 Conclusions 18
References 19

2 How Financial Literacy Can Be Measured? 23
2.1 Introduction 23
2.2 Big Three and Big Five 24
2.3 OECD/INFE Toolkits: Individuals and Entrepreneurs 27
2.4 Measurement Error and Endogeneity Issues 30
2.5 Critical Perspectives on Financial Literacy and Further Research Approaches 32
2.6 Conclusions 36
References 36

3 Financial Literacy and Related Financial Outcomes 41
3.1 Introduction 42
3.2 Savings and Financial Resilience 42
3.3 Access to Credit 44
3.4 Investment Behavior and Decisions 46
3.5 Retirement Planning 48
3.6 Insurance 49
3.7 Conclusions 51
References 58

4 Financial Literacy and Current Challenges: Digitalization and Sustainability 63
4.1 Introduction 64
4.2 Financial Literacy, Digitalization, and Sustainability: A Literature Review 65
4.3 Financial Literacy and Fintech Adoption 67
4.4 Financial Literacy and Sustainable Finance Decisions 73
4.5 Conclusions 80
References 81

Final Remarks 85

Index 89

About the Author

Edoardo Lanciano is a Post-Doc in Finance and Teaching Assistant at the University of Roma Tre, Italy. He gained a PhD in Banking and Finance at the University of Rome Tor Vergata, Italy. His research interests and publications are mainly focused on the topics of financial literacy, financial education, and financial decision-making.

Abbreviations

ATM	Automated Teller Machine
BRICS	Brazil, Russia, India, China, South Africa
ESG	Environmental, Social and Governance
EU	European Union
GDP	Gross Domestic Product
INFE	International Network of Financial Education
IV	Instrumental Variable
MSMEs	Micro, Small and Medium sized Enterprises
NRRP	National Recovery and Resilience Plan
OECD	Organization of Economic Cooperation and Development
OLS	Ordinary Least Square
PISA	Programme for International Student Assessment
POS	Point of Sale
SRI	Socially Responsible Investments

List of Figures

Fig. 1.1 Variation in financial literacy around the world (% of financially literate adults) (*Source* Author's elaboration of data from the Global Financial Literacy Survey) 7

Fig. 1.2 Variation in financial literacy in Europe, Asia, and Latin America (financial literacy scores) (*Source* Author's elaboration of data from the international survey of adult financial literacy) 9

Fig. 1.3 Mean financial literacy scores in PISA 2022 (*Source* Author's elaboration of data from the PISA 2022 survey) 12

List of Tables

Table 3.1	Bibliographic sources on financial literacy and financial outcomes	52
Table 4.1	Fintech adoption questions	69
Table 4.2	Financial literacy and fintech adoption in adult people	70
Table 4.3	Financial literacy questions (Bank of Italy survey on young people's financial literacy)	72
Table 4.4	Financial literacy and fintech adoption in young people	74
Table 4.5	Knowledge of sustainable finance concepts	75
Table 4.6	Financial literacy and knowledge of sustainable finance concepts	76
Table 4.7	Financial literacy summary statistics	77
Table 4.8	Impact investment questions	78
Table 4.9	Financial literacy and impact investment decisions	79

CHAPTER 1

Financial Literacy and Financial Education: An Overview

Abstract This first chapter aims to introduce the topic of financial literacy by providing and explaining the main definitions and analyzing the meaning and the different components of financial literacy. Subsequently, the chapter reports general evidence of the level of financial literacy all around the world, available from international survey data such as OECD and World Bank surveys, and descriptive analyses of the distribution of financial literacy among different sociodemographic categories, showing gaps related to gender, education, geographical origin, and income. Finally, the chapter provides an overview of the actors with key roles for the general development of financial literacy in society, such as governments, supervisory authorities, educational institutions, and financial intermediaries, analyzing their contribution in spreading financial education in the society.

Keyword Definition · Survey · Sociodemographic categories · Society

E. Lanciano, *The Relevance of Financial Literacy*,
https://doi.org/10.1007/978-3-031-93397-4_1

1.1 Introduction

The knowledge and understanding of the topic of financial literacy certainly begin with an examination of the definition of the concept, which appears to be quite complex and articulated. In fact, financial literacy is a concept that touches different aspects of individual knowledge and behavior. It is also important to distinguish literacy from education activities, concepts that are often confused and used in place of each other. The examination of the definitions ends with the explanation of the main financial concepts to be known for being considered as basically financially literate.

The next step in starting to understand the economic and social relevance of financial literacy globally is to analyze the spread of financial literacy at the global level. The analysis is conducted through the study of evidence emerging from international survey experiences. Specifically, this chapter will describe the evidence from three international surveys: the Global Financial Literacy Survey conducted by Standard & Poor's Ratings Services and the World Bank, the International Survey of Adult Financial Literacy conducted by OECD, and the PISA[1] survey on young people conducted again by OECD. The surveys show an overall low level of financial literacy and significant disparities between different geographic areas and between sociodemographic groups. At the geographical level, developed countries have higher levels of financial literacy than emerging countries, while at the sociodemographic level, specific groups appear to be more vulnerable, such as women, the young and elderly, the less educated, and the less wealthy.

Finally, for a comprehensive understanding of the topic, the chapter describes who are the key players in the policy and regulatory spheres involved in financial literacy, who set long-term strategies and goals to increase the level of financial literacy of the population and thus improve the overall status of well-being. These actors include from governments or public authorities to companies, associations, or private environments such as families.

[1] Programme for International Student Assessment. https://www.oecd.org/en/about/programmes/pisa.html#:~:text=PISA%20is%20the%20OECD%27s%20Programme%20for%20International%20Student,science%20knowledge%20and%20skills%20to%20meet%20real-life%20challenges.

1.2 Definitions

Financial literacy is a concept with a complex nature, which is often simplified and approximated to the financial knowledge, while indeed it is characterized by various other factors. The definition that is certainly most widespread and adopted is the one provided by OECD[2] (2024b), which defines financial literacy as *"a combination of financial awareness, knowledge, skills, attitudes and behaviors necessary to make sound financial decisions and ultimately achieve individual financial well-being"*. Thus, it is possible to deduct that it is not a matter of knowledge only, but there are also other elements related to the behavioral, decisional, and psychological fields, such as awareness, skills, and attitudes in making financial decisions. In addition, the definition includes also specific purposes, such as making aware, and possibly correct, financial decisions and reaching an individual financial well-being.

Regarding this last aspect, more precise purposes of owning and adequate level of financial literacy are included in the definition provided by Lusardi and Mitchell (2014), which describe financial literacy as the *"peoples' ability to process economic information and make informed decisions about financial planning, wealth accumulation, debt, and pensions"*, and thus referring to specific kinds of financial decisions, related to specific areas such as savings, investments, retirement planning or liability management. In this definition too, the complexity of the topic emerges. In fact, the ability of processing economic information includes not only knowledge but also the behavioral and psychological factors for making financial decisions and, moreover, this definition illustrates the variety of financial decisions related to everyday people's life for which this ability is needed.

Frequently, financial literacy is confused with financial education. The two expressions are, in fact, used very often with the same meaning, while there are relevant differences between them. The OECD (2024c) indeed defines financial education as *"the process by which financial consumers/investors improve their understanding of financial products, concepts and*

[2] The Organization of Economic Cooperation and Development conducts research, provides policy analysis, and facilitates international cooperation to support policymakers and other stakeholders in developing effective financial consumer protection, financial education, and financial inclusion policies. OECD governments officially recognized the importance of financial literacy in 2002 by launching a unique and comprehensive project to support financial literacy globally.

risks and, through information, instruction and/or objective advice, develop the skills and confidence to become more aware of financial risks and opportunities, to make informed choices, to know where to go for help, and to take other effective actions to improve their financial well-being". Financial education too thus goes beyond the provision of financial knowledge and refers also to behavioral and attitudinal aspects. The different meaning between financial literacy and education emerges by comparing the two definitions. Financial literacy, being represented as the result of a combination of various elements, has a static nature and is considered as the consequence of owning adequate financial knowledge, skills, and attitudes. Financial education, instead, is represented as a process, thus providing the idea of a dynamic nature. Specifically, is it the process through which an individual aims to increase his level if financial literacy, by improving the knowledge and the understanding of financial concepts and products, and developing skills and confidence on financial decision-making. It is possible to say that financial literacy is an outcome, while financial education is a tool, and specifically the tool for improving financial literacy. Indeed, initiatives and programs aimed at spreading and increasing the level of financial literacy among the population are usually classified as financial education initiatives and programs.

It is also interesting to discuss about the topics, e.g., the financial concepts. Which concepts do we refer to when talking about financial literacy? Using the term literacy, means providing basic and essential elements in order to spread necessary and elementary skills and capabilities. Thus, being financially literate does not necessarily mean to be an expert or a technician in finance, but it means to have an adequate understanding of key financial concepts required for managing personal finances, such as interest rates, inflation, risk diversification, insurance or longevity risk, which, for example, allow people to understand that money have a cost, that prices may change according to economic variables, that investment decisions are characterized by different risk-return combinations, or that it is important to protect against risks and to think in time about the future. These topics will be deepened in the sections related to the different measures of financial literacy and the relationships between financial literacy and financial outcomes.

1.3 Evidence from Around the World: Survey Experiences

A first step for understanding the relevance of financial literacy for the society and for developing effective financial literacy policies, goals and strategies, and financial education programs, is to know the level of financial literacy of the population and how it is distributed worldwide. Indeed, collecting reliable evidence and data internationally is fundamental for understanding, for example, social differences in this field based on different geographical areas or among different groups of people, or also which financial concepts and skills need more improvements compared to other ones, beyond the overall level of financial literacy worldwide.

The investigations conducted at global level highlight how the level of financial literacy around the world is generally low and changes significantly among different geographical areas. In this section, the most relevant survey experiences aiming at investigating the level of financial literacy globally will be discussed and analyzed: the Global Financial Literacy Survey conducted by Standard & Poor's Ratings Services and the World Bank[3] (Lusardi et al. 2015), the latest edition of the International Survey of Adult Financial Literacy conducted by OECD[4] (2023a), and finally the Programme for International Student Assessment (PISA) survey conducted by OECD in 2022 (OECD 2024a).

The Global Financial Literacy Survey represents the largest global measurement of financial literacy. More than 150,000 randomly selected individuals in more than 140 countries, aged from 15 and over, were involved. The respondents were interviewed face-to-face or by phone. Data were collected in 2014. In this survey, financial literacy was measured through four questions aiming at investigating basic knowledge and understanding of four key financial concepts: interest rates, interest compounding, inflation, and risk diversification. The respondent was considered as financially literate with three correct answers of these four topics. The first evidence emerging from this investigation is that only 33% of adults worldwide are financially literate, immediately showing the overall low level of financial literacy of the global population. This

[3] https://gflec.org/initiatives/sp-global-finlit-survey/.

[4] https://www.oecd.org/en/publications/oecd-infe-2023-international-survey-of-adult-financial-literacy_56003a32-en.html.

means that only one in three adults shows a basic understanding of essential financial concepts, and thus that around 3.5 billion adults globally lack this basic understanding (Lusardi et al. 2015).

Once considered the level of financial literacy of the overall population, it is interesting to analyze the relevant differences that emerged among countries and geographical areas. The average country-level financial literacy ranges between 13% and 71% of financially literate individuals. From the investigation is immediately evident a relevant gap between countries with well-developed financial markets and emerging countries. Several western countries, such as USA, Canada, United Kingdom, northern European countries, Australia included, represent the geographical areas that report higher levels of financial literacy, with percentages of financially literate adults between 55 and 71%. The 10 countries with the highest percentages of financially literate adults are Sweden, Denmark and Norway (71%), Israel, Canada (68%), United Kingdom (67%), the Netherlands, Germany (66%), Australia (64%), and Finland (63%). On the other hand, lower levels of financial literacy are recorded in South America, Africa, Middle East, and South Asia with many countries with less than 34% of financially literate respondents. Compared to the major western economies, in the most representative emerging countries, such as the so-called BRICS (Brazil, Russia, India, China, and South Africa), on average 28% of adults are financially literate. The average difference between developed and emerging countries in the level of financial literacy can be clearly seen in the following graph, which reports and compares the percentages of financially literate people in the seven most advanced economies (Canada, France, Germany, Italy, Japan, United Kingdom, and United States in blue) and in the major emerging economies (Brazil, China, India, Russia, and South Africa in orange), taken from the Global Financial Literacy Survey (Fig. 1.1).

Focusing on Europe, the average percentage of financially literate individuals is 52% but with several differences among European countries. Northern countries, for example, report better financial literacy scores compared to the rest of the continent. Denmark, Germany, the Netherlands, and Sweden have the highest percentages in Europe with at least 65% of financially literate adults. These rates are significantly lower, and below the average European level, in countries from southern Europe such as Greece (45%), Spain (49%), Italy (37%), and Portugal (26%). Financial literacy rates are low on average also for eastern Europe

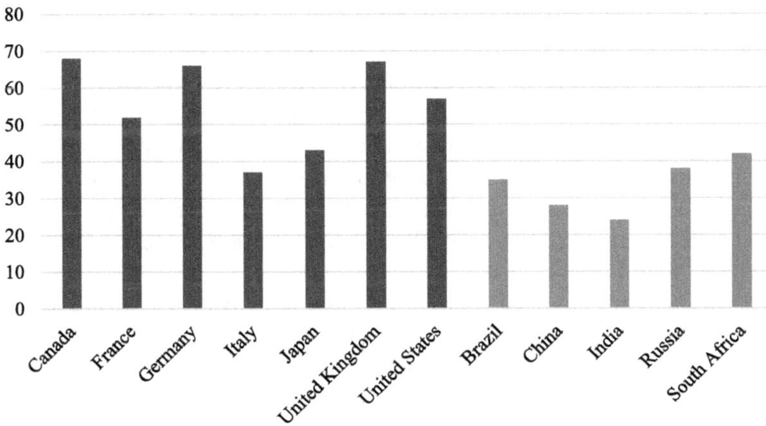

Fig. 1.1 Variation in financial literacy around the world (% of financially literate adults) (*Source* Author's elaboration of data from the Global Financial Literacy Survey)

countries such as Bulgaria (35%) and Romania (22%) with this last one reporting the lowest rate in Europe.

Considering this evidence, financial literacy is usually higher in richer countries, and thus it is possible to assume that the level of financial literacy reflects the overall level of economic and social development. Indeed, financial literacy significantly changes between developed and emerging countries. Probably, two global maps representing respectively the level of financial literacy of a country and the amount of Gross Domestic Product (GDP) per capita at country level should be very similar.

Regarding the topics used in the survey for assessing the individual level of financial literacy, interest rate, and inflation are the topics that report the higher percentages of correct answers, with half of the adults worldwide on average that show a basic understanding of these concepts, while risk diversification is less understood, with only the 35% of respondents correctly answering to this question. Finally, the Global Financial Literacy Survey highlights relevant differences not only between countries and wider geographical areas but also among different sociodemographic groups. For example, women, less educated, and less wealthy respondents record lower financial literacy scores, and these differences are often reported for both developed and emerging countries. This aspect will be deepened in the next section on sociodemographic differences.

The latest edition of the International Survey of Adult Financial Literacy conducted by OECD (2023a) shows overall similar evidence. This survey was conducted by OECD in 2023, after two previous editions in 2020 and 2017. This investigation included 39 countries from Europe, Asia, and Latin America, of which 20 are OECD member countries and 8 are G20 members, for a total amount of 68,826 respondents aged between 18 and 79.

In this survey, financial literacy was assessed using the 2022 OECD/INFE Toolkit for Measuring Financial Literacy and Financial Inclusion (OECD 2022). According to this toolkit, financial literacy is computed through a set of questions covering several aspects of three components that OECD considers in defining financial literacy: financial knowledge, financial behavior, and financial attitudes. Financial knowledge is considered as a basic understanding and the ability to apply basic numeracy skills to essential financial concepts such as interest rate, inflation, risk diversification, and risk-return relationship, and it is measured through seven different questions forming a score between 0 and 7. Financial behavior is assessed through a set of questions aiming at investigating individuals' actions related to their finances such as budgeting, planning and managing finances. More specifically, these questions ask the respondents about payment habits, long-term planning, savings, and purchasing decisions. The financial behavior score ranges between 0 and 9. Finally, financial attitude investigates individual attitudes towards money and preferences for spending money in the short term. The financial attitude questions form a score between 0 and 4. The overall financial literacy score is the sum of correct answers to the financial knowledge, financial behavior, and financial attitude questions, and thus ranges between 0 and 20. The OECD sets 70% of correct answers as a minimum target to this score for considering an individual as essentially financially literate. This means that with 70% of this score an individual has a basic understanding of financial concepts and is able to apply prudent principles on their personal finances and in dealing with the financial sector.

Going to the survey results, the evidence is very similar and consistent with the previous ones obtained in the Global Financial Literacy Survey from Standard & Poor's and World Bank. In this case too, the level of financial literacy worldwide seems to be quite low. Indeed, the average financial literacy score obtained by the participating countries is

60%, below the minimum target, with better average score obtained in knowledge (63%) rather than behavior (61%) and attitude (56%). Considering the overall sample, only 34% of the respondents reach the minimum target score of financial literacy. This evidence is very similar to the Global Financial Literacy Survey, where the percentage of financially literate adults worldwide was 33%.

The OECD survey too reports relevant heterogeneity at the geographical level. European countries obtained the higher average financial literacy score (62%) and were the most numerous among the 10 countries with the highest percentages of financial literacy, with Germany, Sweden, Finland, Ireland, Estonia, and Malta. Lower average scores were obtained by Asian (59%) and southern American countries (56%), with these last ones with the most representatives (Paraguay, Panama, Chile, Costa Rica, Peru) among the 10 countries with the worst scores. The difference between Europe, Asia, and Latin America emerging in this survey can be seen in the following graph (Fig. 1.2), which represents the differences in financial literacy scores obtained by the five best-performing countries in Europe (in blue), Asia (in yellow), and Latin America (in red).

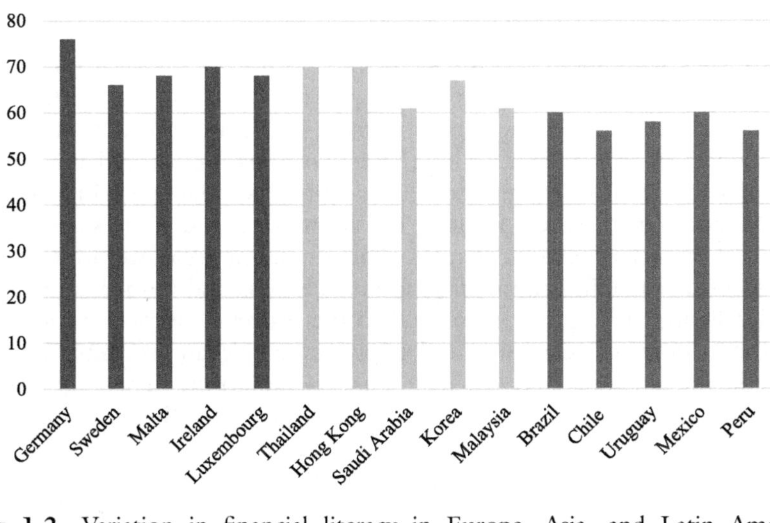

Fig. 1.2 Variation in financial literacy in Europe, Asia, and Latin America (financial literacy scores) (*Source* Author's elaboration of data from the international survey of adult financial literacy)

The average financial literacy scores obtained by these five European countries (Germany, Sweden, Malta, Ireland, and Luxembourg) is 70%, while this percentage is 66% for the five Asian countries (Thailand, Hong Kong, Saudi Arabia, Korea, and Malaysia) and 58% for the five Latin American countries (Brazil, Chile, Uruguay, Mexico, and Peru).

Finally, the 2022 PISA survey, developed again by OECD, measures knowledge and skills of 15-year-old students in a set of disciplines such as mathematics, science, or reading, including also financial literacy. The PISA 2022 assessment of financial literacy is the fourth of its kind. Moreover, financial literacy performances are linked to various sociodemographic factors. This investigation involves 20 participating countries, whose 14 are OECD members (Austria, Belgium, Canada, Costa Rica, Czechia, Denmark, Hungary, Italy, the Netherlands, Norway, Poland, Portugal, Spain, and the United States) and 6 are partner countries (Brazil, Bulgaria, Malaysia, Peru, Saudi Arabia, and the United Arab Emirates), and around 98,000 students were part of this assessment, representing about 9.5 million 15-year-olds students in the schools of the 20 participating countries.

In this survey, financial literacy assessment consists of a wide set of questions, composed according to the PISA 2022 assessment and analytical framework (OECD 2023b), asking the students about their experiences with money matters, their exposure to financial education courses in school, their ability to apply the financial knowledge and skills that they have learned in or outside of school, financial attitudes, and behaviors. More specifically, these questions are divided into three sections: content areas, processes, and contexts. The questions related to the content areas are focused on knowledge and understanding of essential financial literacy concepts. These content areas are: money and transactions, planning and managing finances, risk and reward, and financial landscape. The questions related to the processes section are focused on their ability to recognize and apply these relevant concepts and to understand, analyze, and evaluate financial solutions. These questions are divided in four categories, which are: identify financial information, analyze financial information and situations, evaluate financial issues, and apply financial knowledge and understanding. Finally, the questions related to the contexts are focused on situation and contextual factors that may be related to students' opportunities to improve their financial literacy. These contexts are: education and work, home and family, individual habits and society.[5]

Average differences in financial literacy between developed and emerging countries, or between western and non-western countries emerge in this survey too. The main results show that Austria, Belgium, Canada, Czechia, Denmark, the Netherlands, and Poland performed above the OECD average in financial literacy. More in detail, for example, more than 15% of students in Belgium and the Netherlands were top performers in financial literacy, compared to less than 1% of students in Malaysia and Saudi Arabia. Additionally, in Brazil, Bulgaria, Costa Rica, Malaysia, Peru, and Saudi Arabia, more than 40% of the respondents performed at or below Level 1,[6] which is the lowest level of difficulty to reach, and means having basic abilities to recognize the difference between needs and wants, making simple decisions about everyday spending, and recognizing the purpose of everyday financial documents, such as an invoice. This percentage is on average 18% across the OECD countries and, for example, 11% in Denmark.

The following graph represents the average results obtained by students from each country in the financial literacy assessment (Fig. 1.3), where the difference between the average score obtained by OECD countries (in green) and partner countries (in orange) immediately emerges.

1.4 Sociodemographic Differences

As mentioned in the previous section, differences in the level of financial literacy occur not only among geographical areas but also between sociodemographic groups. Indeed, the level of financial literacy is absolutely non-homogeneous and changes significantly according to specific sociodemographic characteristics such as gender, age, educational level,

[5] For further details on the PISA financial literacy questions please consult the PISA 2022 assessment and analytical framework (OECD 2023).

[6] PISA scales are divided into proficiency levels. For example, for PISA 2022, the range of difficulty of financial literacy items is represented by five levels of financial literacy proficiency: the simplest items correspond to Level 1, while Levels 2, 3, 4, and 5 correspond to increasingly difficult items.

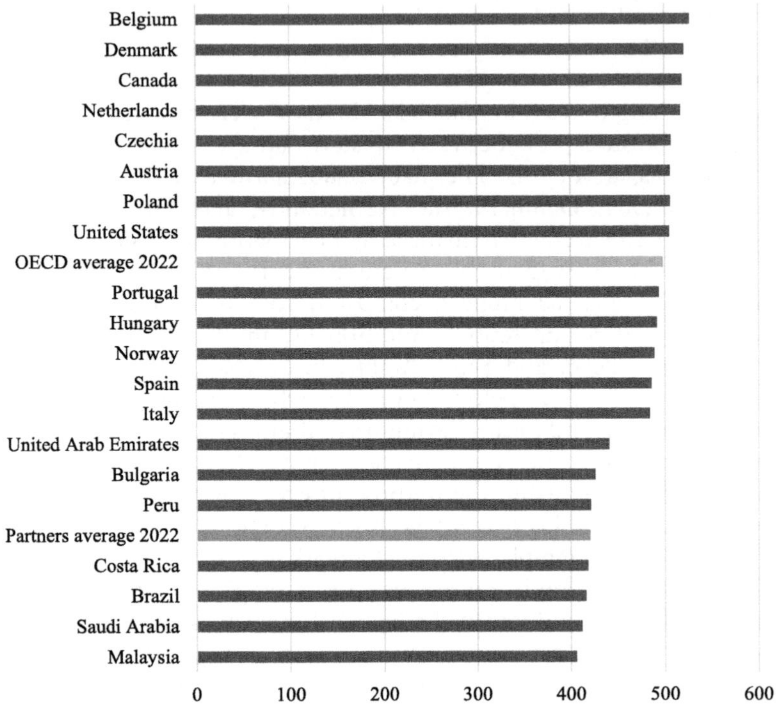

Fig. 1.3 Mean financial literacy scores in PISA 2022 (*Source* Author's elaboration of data from the PISA 2022 survey[7])

or income. The literature illustrates that some sociodemographic characteristics are highly correlated with financial literacy (Muñoz-Murillo et al. 2020).

Financial literacy rates differ in a relevant way according to gender. Men tend to have significantly higher financial literacy than women, and this difference occurs all around the world, in several investigations, both in developed and emerging countries. Gender differences in financial literacy are a thus global phenomenon. Worldwide, 35% of men are financially literate, while for women this percentage is 30%, with some contexts

[7] For further details on the PISA financial literacy scores please consult the PISA 2022 assessment and analytical framework (OECD 2023).

where this difference reaches 15 percentage points (Klapper and Lusardi 2020). Bottazzi and Lusardi (2021) show that gender strongly correlates with financial literacy as women generally report a lower level than men in knowledge and understanding of financial matters. This also explains part of the difference between men and women in the participation of financial markets (Almenberg and Dreber 2015). Greater participation of financial markets by men also leads to investing more in learning about these markets and financial concepts than women, which historically have been socialized to focus more on family issues (Driva et al. 2016). These phenomena are changing, but only slowly. Not only adult men are generally more financially literate than adult women but similar findings also emerge from younger people as well (Lusardi et al. 2010; Chen and Volpe 2002). Young males perform better than girls in several countries (Austria, Costa Rica, Denmark, Hungary, Italy, and Portugal) according to the PISA 2022 results (OECD 2024).

Young people represent another vulnerable group in financial literacy. According to the results of global investigations, financial literacy is lower among young people across several countries and economies. The Global Financial Literacy Survey (Lusardi et al. 2015) shows that in major advanced economies, on average, 56% of the respondents aged 35 or younger are financially literate, against 63% of the respondents aged between 36 and 50.

The overall lack of financial literacy among young people can be explained by various socio-economic factors, such as difficulties in creating an own and stable working life, especially in certain countries, after several economic and occupational crises in the recent years, which does not allow to reach financial independence in an early age. For this reason, there is an overall consensus among financial education experts that schools are one of the most appropriate venues to deliver financial education initiatives, and that teaching these concepts to young people, especially during the schooling period, can be crucial in terms of long-term goals (Lusardi et al. 2010; Kaiser and Menkhoff 2020). Financial literacy is also generally lower among the old. This evidence is persistent in many countries (Lusardi and Mitchell 2011a, b). According to the evidence from the Global Financial Literacy Survey, financial literacy rates are lower for adults older than 50, and the lowest among those older than 65. Moreover, older people often tend to overestimate their knowledge and skills on financial matters, declaring to possess high levels of self-assessed financial literacy while actually scoring poorly on basic financial literacy questions (Lusardi and Mitchell 2011a; Lusardi and Tufano 2009).

Financial literacy is also higher among well-educated and wealthy people. This evidence also is persistent across several countries and investigations. On average, people with university level education obtain better scores in financial literacy compared to people with high school or lower education. Education certainly may improve numeracy and text comprehension skills, and cognitive abilities, all relevant factors for making aware and possibly correct financial decisions (McArdle et al. 2009). In addition, various studies report the relationship between financial literacy on young people and the education of their parents, finding a significant and positive association between financial literacy and the parental educational background (Lusardi et al. 2010; Mahnaz and Horton 2012). Financial literacy usually changes according to the level of income. Income is strongly associated with financial literacy, and financial literacy levels are usually higher in people with higher incomes compared to people with lower incomes. Clearly, a higher income allows to have more possibilities to manage financial resources, or to make long-term economic planning or investment decisions, and thus in general to deal more with financial matters. On the contrary, knowing basic financial concepts would be especially important for people with low and middle incomes, who are often struggling to make ends meet and are less likely to invest in this knowledge (Buckland et al. 2024; French and McKillop 2016). Lower income forces people to focus more on the present and therefore to take their focus away from long-term planning. Income is obviously also highly related with employment. Indeed, unemployed people or out of the labor force show lower levels of financial literacy (OECD 2023a). Differences in financial literacy related to wealth also emerge in the 2022 PISA survey (OECD 2024a), which shows that students having disadvantaged socioeconomic backgrounds perform lower in financial literacy, while more advantaged students usually performed better.

Finally, other differences between sociodemographic groups are reported in the existing literature. Indeed there are differences between people living in rural areas and people living in the cities (Klapper and Panos 2011), or internal geographical differences in certain countries such as Italy, with northern people obtaining higher scores than southern people (Fornero and Monticone 2011), or Romania (Beckmann 2013), or United States (Bumcrot et al. 2013), or also differences related to ethnicity, with for example African Americans and Hispanics showing lower levels of financial literacy in the United States (Lusardi and Mitchell 2007a, b, 2011a).

1.5 Key Players

Financial literacy, especially in the last decade, has acquired an increasing relevance as a policy goal. Indeed, considering the overall low level of financial literacy in many areas of the globe, it has acquired a growing interest among policymakers, financial authorities, or academics (Nicolini and Cude 2021), and it has become a long-term policy goal in many countries and economies, and multiple actors and stakeholders participate to this process of development and improvement of financial literacy for individuals and society, from governments, to public and private entities. In order to spread financial literacy among the population, it is actually important to develop synergies and collaborations between policymakers, supervisory authorities, financial institutions, and the educational sector. In recent years, many countries have adopted their own national strategies for financial literacy, aiming at promoting and coordinating initiatives, training, and development programs (Is et al. 2024), and at identifying good practices and methodologies for increasing the overall level of financial literacy. These national strategies define a roadmap for empowering financial consumers and investors and follow specific principles and guidelines developed by OECD (2012). International organizations such as OECD or World Bank cover indeed a crucial role for financial literacy by developing guidelines and recommendations, promoting initiatives, and conducting global investigations. According to these principles, the national strategies should be guided by a leading public authority and should involve multiple stakeholders. This leading public authority could be, for example, a government, a ministry (in particular the ministries of finance or education), or a regulatory and supervisory authority such as a central bank, or also a new dedicated body such as a national council or committee.[8]

The leading authority sets the overall strategy framework, defines the priorities and goals of the strategies and the related time period, the governance mechanisms, the quality standards of the initiatives, and coordinates the activities of the other participant stakeholders. Among the other involved stakeholders, certainly, the private sector and financial services providers should be included. The participation of private actors is necessary, considering their expertise and knowledge of the market, and for their possibility to reach a wide target audience, but at the same time it should be monitored in order to avoid risks of potential conflicts of interests or promotion of their financial products or services. In national strategies, private actors and financial service providers cover a relevant role in the implementation of financial education initiatives, training programs, or in the provision of dedicated material, or in partnerships with public actors.

Other stakeholders that may be involved in the national strategies are represented by local organizations and civil society actors such as, for example, research centers, non-governmental organizations, trade unions, consumer associations, media, or other national disseminators (OECD 2012; Buckland 2018). The role of local organizations and civil society can be important, especially concerning the organization and the implementation of initiatives at the local level, in regions, provinces, or even in small municipalities.

Concrete examples of actors involved in national strategies for financial literacy are provided by OECD, which conducted in 2021 an investigation aiming at assessing the national strategies of 29 countries, which analyzes information gathered through a survey submitted in 2019 and 2020[9] (OECD 2022). This investigation is particularly focused on the

[8] In Italy, for example, the main body committed in defining the national strategy for financial literacy is the Committee for the planning and coordination of financial education activities (Edufin Committee). This body, established in 2017 by the Ministry of Economy and Finance, aims to promote and coordinate initiatives at national level to improve financial, insurance, and pension knowledge and skills among the population. https://www.quellocheconta.gov.it/it/.

[9] This Evaluation of National Strategies for Financial Literacy summarizes and draws lessons from the evaluation approaches and activities of 29 countries and economies. It provides implementation guidance on the evaluation of their national strategies for financial literacy, accompanied by concrete examples gathered through a series of

leading actors of the strategies. The authority that most often appears as the leading authority of the national strategy is the central bank (Austria, Germany, Spain, Turkey, Japan, India, Portugal, Hungary). Other authorities that frequently appear as the leading ones are governments and specifically ministries of finance (Czech Republic, Hungary, Netherlands, Poland), or also financial markets supervisory authorities (United Kingdom, United States, Canada, Spain, Chile, Perú, Paraguay, Indonesia). In some cases (Spain, Hungary), the strategy is jointly guided by multiple authorities. The national strategies analyzed in this investigation often involve various stakeholders such as other public actors, national committees, and are often based on a public–private partnership model (OECD 2022).

It is also necessary to mention the crucial role of the educational sector, including schools and universities. There is indeed a consensus among financial literacy and education experts that schools and universities are probably the most appropriate venues to deliver financial education initiatives, and that teaching these concepts to young people, especially during the schooling period, can be crucial in terms of long-term goals (Lusardi et al. 2010; Kaiser and Menkhoff 2020). Most of the financial education initiatives are actually directed to young people in an educational context, and in many cases, these initiatives are particularly effective. Studies on high school and university students are significantly more widespread than studies on younger students (Mandell 2008; Boyce and Danes 1998; Amagir et al. 2018; Frisancho 2018; Lusardi et al. 2018; Lührman et al. 2018). Kaiser and Menkhoff (2020), in a literature review of studies on the effectiveness of school-based financial education programs in Europe, USA, South America, and Africa, show that, on average, financial education treatments have a positive and significant impact on the financial literacy of students in different educational cycles. A similar work has been conducted by Amagir et al. (2018) who, through a systematic literature review, assessed the effectiveness of 36 financial education programs aimed at children and adolescents distributed across Europe, USA, and Africa, and also try to understand what the typical characteristics of an

interviews and a survey conducted in 2019 and 2020. https://www.oecd.org/en/publications/evaluation-of-national-strategies-for-financial-literacy_91e310db-en.html#:~:text=This%20Evaluation%20of%20National%20Strategies%20for%20Financial%20Literacy,the%20OECD%20International%20Network%20for%20Financial%20Education%20%28OECD%2FINFE%29.

effective program are. Evidence from these studies shows that school-based programs are generally able to improve students' level of financial literacy. A further work by Bernheim et al. (2001) highlights, through a survey, how among adults from United States who have taken a financial education course at a young age, greater amounts of savings are recorded.

Finally, a relevant role can be covered also by the context where individuals grow and learn their first consumer and money management habits, represented by the family and the life context. Families are important especially considering the difference, on average, between the level of financial literacy of adult and young people. Individual financial literacy can be affected, in positive or in negative way, by familiar context (Lusardi 2015). According to various studies, indeed, financial literacy can begin from the family, where children can observe their parents' saving, consuming, and investing habits, or receiving financial education directly from parents (Chiteji and Stafford 1999; Li 2009; Shim et al. 2009). The PISA 2022 survey results (OECD 2024a) show how students who discuss with their parents about their saving or purchasing decisions are, on average, much more financially literate, compared to those who reported to never discuss these issues. The role of families, based especially on discussions about financial resources management (Hanson and Olson 2018), seems thus to be fundamental for a young individual. Regarding the life context, for example, low-income neighborhoods are generally less exposed to opportunities for using financial services (Buckland 2012).

For the overall development of financial literacy among the population, it is thus important that the participation and the cooperation of several key players, from governments, ministries, financial authorities, dedicated committees, to private actors, financial intermediaries, associations, and finally families and local communities.

1.6 Conclusions

The complex definition of financial literacy, characterized by multiple elements, makes immediately understandable the multifaceted nature of the concept. It is also important that the difference between financial literacy and financial education is clear so that these concepts or the terminology will not be confused, and it is useful to remember that financial literacy has to do with simple concepts, essential in the personal finance management, and not with concepts for experts or with a high level of difficulty.

The global surveys analyzed show similar results among each other, both at the geographical and sociodemographic levels, and generally tend to show that overall levels of financial literacy are quite low, especially in some sociodemographic groups, causing social inequalities.

For these reasons, financial literacy has become an increasingly important issue in the political agenda of many countries, where national strategies for financial literacy have been defined, in which even the highest political bodies, such as governments, ministries, or central banks have a key role.

The main purpose of this chapter is to provide the reader with a learning basis of the concept, through the explanation of definitions, concepts, the description of the global evidence, and the actors involved in policy and regulatory decisions, in order to then be able to better understand the content of the following chapters, where aspects of the research on the topic and the relevance of the phenomenon will be explored more in depth.

References

Almenberg, J. and A. Dreber (2015). Gender, stock market participation and financial literacy. Economics Letters 137, 140–142.

Amagir, A., Groot, W., Van Den Brink, H., Wilschut, A. (2018). A review of financial literacy education programs for children and adolescents. Citizenship, Social and Economics Education 2018, Vol. 17(1) 56–80.

Beckmann, E. (2013). "Financial Literacy and Household Savings in Romania." Numeracy 6 (2).

Bernheim, B. D., Garrett, D. M., Maki, D. M. (2001) Education and saving: the long-term effects of high school financial curriculum mandates, Journal of Public Economics, 80, 435–65.

Bottazzi, L. and A. Lusardi (2021). Stereotypes in financial literacy: Evidence from pisa. Journal of Corporate Finance 71, 101831.

Boyce, L. and Danes, S. (1998) Evaluation of the NEFE High School Financial Planning Program, National Endowment for Financial Education (NEFE), Engelwood, CO.

Buckland, J. (2012). Hard Choices: Financial Exclusion, Fringe Banks and Poverty in Urban Canada. University of Toronto Press. https://doi.org/10.3138/9781442685291.

Buckland, J. (2018). Building Financial Resilience: Do Credit and Finance Schemes Serve or Impoverish Vulnerable People? Springer International Publishing, pages 149–188.

Buckland, J., Frimpong, J., Nur, W., & Simpson, W. (2024). Financial capability of people with low income: Results from the Canadian financial diaries. Poverty & Public Policy, 16(1), 8–23. https://doi.org/10.1002/pop4.394.

Bumcrot, C., Lin, J., Lusardi, A. (2013). "The Geography of Financial Literacy." Numeracy 6 (2).

Chen, H. and R. P. Volpe (2002). Gender differences in personal financial literacy among college students. Financial services review 11(3), 289–307.

Chiteji, N.S., Stafford, F. (1999). "Portfolio Choices of Parents and Their Children as Young Adults: Asset Accumulation By African-American Families." American Economic Review 89 (2): 377–80.

Driva, A., Lührmann, M., & Winter, J. (2016). Gender differences and stereotypes in financial literacy: Off to an early start. Economics Letters, 146, 143–146. https://doi.org/10.1016/j.econlet.2016.07.029.

Finke, Michael S., John Howe, and Sandra J. Huston. 2011. "Old Age and the Decline in Financial Literacy." SSRN http://ssrn.com/abstract=1948627.

Fornero, E., Monticone, C. (2011). "Financial Literacy and Pension Plan Participation in Italy." Journal of Pension Economics and Finance 10 (4): 547–64.

French, D., McKillop, D. (2016). Financial Literacy and Over-Indebtedness in Low-Income Households. International Review of Financial Analysis 48: 1–11. https://doi.org/10.1016/j.irfa.2016.08.004.

Frisancho, V. (2018). The impact of school-based financial education on high school students and their teachers: Experimental evidence from Peru. Inter-American Development Bank Working Paper No. 871.

Hanson, T., Olson, P. (2018). Financial Literacy and Family Communication Patterns. Journal of Behavioral and Experimental Finance, vol. 19, pp. 64–71.

Is, R., Kv, S. & Hungund, S. (2024) MSME/SME Financial Literacy: A Systematic Literature Review and Bibliometric Analysis. Journal of the Knowledge Economy.

Kaiser T., Menkhoff L. (2020), Financial education in schools: A meta-analysis of experimental studies, Economics of education review (78).

Klapper, L., Lusardi, A. (2020). Financial literacy and financial resilience: Evidence from around the world. Financial Management 49(3), 589–614.

Klapper, L., Panos, G. (2011). "Financial Literacy and Retirement Planning: The Russian Case." Journal of Pension Economics and Finance 10 (4): 599–618.

Li, G. (2009). "Information Sharing and Stock Market Participation: Evidence from Extended Families." Board of Governors of the Federal Reserve System Finance and Economics Discussion Paper 2009–47.

Lührmann, M., Serra-Garcia, M., & Winter, J. (2018). The impact of financial education on adolescents' intertemporal choices. American Economic Journal: Economic Policy, 10(3), 309 332.

Lusardi A., Klapper L., Van Oudheusden P. (2015) Financial Literacy around the world: insights from the Standard & Poor's ratings services global financial literacy. Global Financial Literacy Excellence Center.

Lusardi, A. (2015). Financial literacy skills for the 21st century: Evidence from PISA. Journal of Consumer Affairs, vol. 49, no. 3, pp. 639–659.

Lusardi, A., Hasler, A., & Davis, H. (2018). An Analysis of Financial Literacy Among Italian Students. www.gflec.org.

Lusardi, A., Mitchell, O. (2007a). "Baby Boomer Retirement Security: The Roles of Planning, Financial Literacy, and Housing Wealth." Journal of Monetary Economics 54 (1): 205–24.

Lusardi, A., Mitchell, O. (2007b). "Financial Literacy and Retirement Preparedness: Evidence and Implications for Financial Education." Business Economics 42 (1): 35–44.

Lusardi, A. Mitchell, O. (2011a). "Financial Literacy around the World: An Overview." Journal of Pension Economics and Finance 10 (4): 497–508.

Lusardi, A., Mitchell, O. (2011b). "Financial Literacy and Planning: Implications for Retirement Well-Being." In Financial Literacy: Implications for Retirement Security and the Financial Marketplace, edited by Olivia S. Mitchell and Annamaria Lusardi, 17–39. Oxford and New York: Oxford University Press.

Lusardi, A., Mitchell, O. (2014). The economic importance of financial literacy: Theory and evidence. Journal of Economic Literature, 52(1), 5–44.

Lusardi, A., O. S. Mitchell, and V. Curto (2010). Financial literacy among the young. Journal of consumer affairs 44(2), 358–380.

Lusardi, A., Tufano, P. (2009). "Debt Literacy, Financial Experiences, and Overindebted ness." National Bureau of Economic Research Working Paper 14808.

Mahnaz, M., Horton, N. (2012). Financial Literacy among Educated Women: Room for Improvement. The Journal of Consumer Affairs, Volume 48, Issue 2, pages 403–417.

Mandell, L. (2008) High school financial literacy, in Overcoming the Savings Slump. How to Increase the Effectiveness of Financial education: a randomized experiment 835 Financial Education and Savings Programs (Ed.) A. Lusardi, University of Chicago Press Chicago, IL, pp. 257–79.

McArdle, J., Smith, J., Willis, R. (2009). "Cognition and Economic Outcomes in the Health and Retirement Survey." National Bureau of Economic Research Working Paper 15266.

Muñoz-Murillo, M., Álvarez-Franco, P., Restrepo-Tobón, D. (2020). The role of cognitive abilities on financial literacy: New experimental evidence, Journal of Behavioral and Experimental Economics, Volume 84, 101482.

Nicolini, G., Cude, B. J. (2021). The Routledge Handbook of Financial Literacy. (2021). United Kingdom: Taylor & Francis.

OECD (2012), OECD/INFE High-level Principles on National Strategies For Financial Education, OECD Publishing, Paris.

OECD (2022), Evaluation of National Strategies for Financial Literacy.

OECD (2023a), "OECD/INFE 2023 International Survey of Adult Financial Literacy", OECD Business and Finance Policy Papers, No. 39, OECD Publishing, Paris.

OECD (2023b), PISA 2022 Assessment and Analytical Framework, PISA, OECD Publishing, Paris.

OECD (2024a), PISA 2022 Results (Volume IV): How Financially Smart Are Students?, PISA, OECD Publishing, Paris.

OECD (2024b), Recommendation of the Council on Financial Literacy.

OECD (2024c), Recommendation of the Council on Principles and Good Practices for Financial Education and Awareness.

OECD/INFE (2022) Toolkit for Measuring Financial Literacy and Financial Inclusion.

Shim, S., Xiao, J., Barber, B., Lyons, A. (2009). "Pathways to Life Success: A Conceptual Model of Financial Well-Being for Young Adults." Journal of Applied Developmental Psychology 30 (6): 708–23.

CHAPTER 2

How Financial Literacy Can Be Measured?

Abstract This second chapter aims mainly to provide an exhaustive overview of research techniques and methodologies used for measuring financial literacy. Financial literacy is indeed measured with various tools, employed in surveys and scientific studies. These tools are analyzed together with their advantages and limitations. In addition, the chapter is focused on typical problems affecting financial literacy measurements and studies such as measurement errors and other endogeneity issues, and discusses the main strategies to solve them, specifically referring to instrumental variables. Finally, critical aspects of traditional research approaches and alternative perspectives are discussed and analyzed.

Keywords Measurement · Questions · Research · Endogeneity

2.1 Introduction

The aim of this chapter is to introduce the main aspects of the research on financial literacy. First of all, the main measurements used in scientific studies worldwide are examined and described. Financial literacy can be measured using various techniques that have been developed and improved over the years. Examining the measurements is the first concrete step in approaching research in this field and makes it even

E. Lanciano, *The Relevance of Financial Literacy*,
https://doi.org/10.1007/978-3-031-93397-4_2

clearer which concepts are used to consider an individual as adequately financially literate.

These measurement techniques adopt different approaches, of a more or less synthetic nature, and include multiple concepts in the assessments. Different measurements have also been developed based on the target analyzed, such as proxies aimed at individuals or entrepreneurs, or that also measure subtopics of financial literacy such as digital financial literacy.

Each indicator also has its advantages and limitations. A typical problem that emerges from financial literacy measurements, which are generally performed using survey data, is the measurement error. At the same time, however, techniques have been defined to limit the effects of these errors and other endogeneity problems, such as the use of instrumental variables, for which multiple examples selected from the literature are reported.

Finally, different schools of thought and research perspectives are compared and discussed. More specifically, in addition to the approaches that are considered as mainstream, other approaches are also described, including, for example, experimental techniques or theories employed in other disciplines such as sociology or psychology.

2.2 Big Three and Big Five

Measuring financial literacy is a complicated task, considering the complexity of the concept. Furthermore, developing shared measures of financial literacy is important for international comparability over the years. There are various measurements of financial literacy, which have evolved over time and were developed through different approaches.

Probably, the most employed measurements in scientific studies globally are the Big Three and Big Five financial literacy questions (Lusardi and Mitchell, 2008, 2011; Hastings et al., 2013). These measurements were created with the idea of developing a simple, concise but relevant indicator. Indeed, these measurements are based on respectively only three and five questions, aiming at testing the basic knowledge and understanding of fundamental financial concepts, essentially relevant for people's everyday financial decisions.

Starting from the Big Three, these questions aim to investigate whether an individual knows and understands these three basic financial concepts: interest rate, inflation, and risk diversification. These questions are the following:

1) Suppose you had $100 in a savings account and the interest rate was 2% per year. After 5 years, how much do you think you would have in the account if you left the money to grow?

   a) More than $102
   b) Exactly $102
   c) Less than $102
   d) Don't know
   e) Prefer not to say

2) Imagine that the interest rate on your savings account was 1% per year and inflation was 2% per year. After 1 year, with the money in this account, would you be able to buy…

   a) More than today
   b) Exactly the same as today
   c) Less than today
   d) Don't know
   e) Prefer not to say

3) Do you think the following statement is true or false? Buying a single company stock usually provides a safer return than a stock mutual fund

   a) True
   b) False
   c) Don't know
   d) Prefer not to say

The Big Five questions include the Big Three with two additional questions, assessing the understanding of the risk-return relationship, specifically the relationship between interest rates and bond prices, and between the length of a mortgage and the overall interest paid over the life of the loan (Van Hove and Ahunov, 2024). These two more questions are the following:

4) If interest rates rise, what will typically happen to bond prices?

   a) They will rise
   b) They will fall
   c) They will stay the same
   d) There is no relationship between bond prices and the interest rate

e) Don't know
f) Prefer not to say

5) A 15-year mortgage typically requires higher monthly payments than a 30-year mortgage, but the total interest paid over the life of the loan will be less.

a) True
b) False
c) Don't know
d) Prefer not to say

The amount of correct responses to these questions represents a financial literacy index. Thus, these questions verify essential concepts useful for the individual orientation in personal finance management. According to various studies available from the literature, performing well on the Big Three or Five is strongly predictive of savvy financial behaviors, such as retirement planning, saving decisions, and debt management (Angrisani et al. 2023). In fact, reflecting on what actually means being able to answer the Big Three or Five correctly, it means understanding the essential concepts of personal finance management. It means knowing, for example, that money has a cost, that prices can change due to economic variables, that investment risk should be mitigated, or that higher returns correspond to higher risks. While, on the contrary, for example, not understanding a simple interest rate calculation, or not knowing price dynamics or not considering the risks related to investments, means that probably an individual is very likely to make wrong financial decisions.

Surely, there are several advantages in using these measurements but also some limitations. Among the advantages, these indicators are simple, concise, comparable, and consider crucial concepts for financial decision-making. Regarding some limitations, the synthetic nature of these indicators may appear as excessive and could entail the risk of non-adequate accuracy of these measurements (Ranyard et al. 2020). Additionally, in some cases probably these questions appear not very flexible or adaptable to circumstances. Indeed, various authors have developed alternative proxies, by adding new questions or modifying one or more of the Big Three or Five. Anderson and Robinson (2022), for example, replaced the mortgage question of the Big Five with a question on long-term savings, or Van Rooij et al. (2012) used five questions for measuring financial literacy, with only two of them taken from the Big Three. Finally, these questions are focused only on aspects related to financial knowledge and do not consider behavioral or attitudinal aspects.

2.3 OECD/INFE Toolkits: Individuals and Entrepreneurs

Another measurement that instead considers three components of financial literacy (knowledge, behavior, and attitude) was developed by OECD/INFE[1] and described in the 2022 OECD/INFE Toolkit for Measuring Financial Literacy and Financial Inclusion (OECD, 2022). The OECD/INFE toolkit is used for assessing the individual level of financial literacy especially for national and international surveys and was revised on various occasions. The first version of the toolkit was developed in 2010, and it was revised and updated three times: in 2015, 2018, and finally in 2022. This toolkit is composed of a questionnaire, designed to reflect the OECD definition of financial literacy, considered as a combination of various elements. More specifically, this measurement of financial literacy is composed of a set of questions divided into three components: financial knowledge, financial behavior, and financial attitudes.

Starting from the dimension of the financial knowledge, this component of the overall financial literacy score is assessed through seven questions aiming at investigating the individual knowledge and understanding of basic financial concepts and the ability to apply basic numeracy skills to these concepts. These seven questions ask about the understanding and the calculation of simple and compounding interest, the definition of inflation and its impact on spending power, and finally about the concepts of risk diversification and the relationship between risk and reward. It is immediately evident how these topics and the topics covered in the Big Three and Big Five questions are very similar. The correct responses to these questions compose a score between 0 and 7.

[1] The International Network on Financial Education brings together public authorities and other stakeholders to share good practices in the design and implementation of financial literacy policies and programs. It collects internationally comparable data, undertakes policy analysis, and develops policy instruments. OECD governments officially recognized the importance of financial literacy in 2002 by launching a unique and comprehensive project to support financial literacy globally. In 2008, the project was expanded through the creation of INFE. INFE members come from over 130 countries and economies, including developed, emerging, and developing economies, and all G20 countries.

The dimension of financial behavior is assessed through 14 questions aiming at verifying whether an individual is involved and behaves correctly for decisions such as budgeting, planning, and managing finances. Specifically, these questions ask about saving behaviors, management of personal or family budget, liabilities management, awareness of financial products and services, holding financial products, determinants of financial decision-making, keeping track of financial operations, long-term financial planning, and purchasing decisions. Correct answers to these questions compose a score between 0 and 9.

Finally, financial attitude is assessed through two questions aiming at verifying individual preferences towards money and attitudes in spending money in the short term. Specifically, these questions ask about preferences for spending or saving money. The correct responses to these questions compose a score between 0 and 4. Thus, considering these three components, the overall individual level of financial literacy is computed as the sum of the scores obtained in each of these three dimensions, which ranges between 0 and 20.[2]

Considering the increasing relevance of the digital innovation in the world of financial services and the growing spread of financial services provided in digital form, the 2022 version of the toolkit includes also a measure of digital financial literacy. This measurement is set in the same way as the "traditional" financial literacy, thus including the three components of knowledge, behavior, and attitude, and aims to capture how an individual is familiar with digital financial services and aware of the risks related to the use of digital tools.

Specifically, digital financial knowledge is measured through three questions about understanding of digital contracts, use of personal data, and understanding of crypto-assets. Digital financial behavior is assessed through two questions asking about sharing PIN codes or passwords of the bank account, sharing personal finance information, and buying financial products online. Finally, digital financial attitude is assessed through one question asking about using public networks for purchasing decisions, checking security of websites, and terms and conditions in case of an online operation.

Beyond the toolkit for individuals, the OECD/INFE also developed a toolkit for measuring financial literacy for entrepreneurs, and specifically

[2] For more information on the methods used to calculate the scores, please refer to the OECD/INFE Toolkit for Measuring Financial Literacy and Financial Inclusion (2022).

for micro, small, and medium entrepreneurs (MSMEs) (OECD 2020). It is interesting to analyze the differences of this measurement with the toolkit for individuals. In fact, entrepreneurial financial literacy has something in common with individual financial literacy, but there are some differences. First of all, it is defined as "*the combination of awareness, knowledge, skills, attitudes and behavior that a potential entrepreneur or an owner or manager of a micro, small or medium sized enterprise should have in order to make effective financial decisions to start a business, run a business, and ultimately ensure its sustainability and growth*". The definition includes precise references to business activities, and this component also emerges in the questions used for assessing financial literacy. This entrepreneurial financial literacy is thus related specifically to business issues rather than personal ones, and it is not meant as the personal financial literacy of the entrepreneur.

For this measurement too, the questions used in the toolkit are designed to assess financial knowledge, behavior, and attitude of a business owner. About the topics, the questions cover financial products used for the business, managing and planning business finances, financial knowledge, and a range of behaviors and attitudes related to financial literacy and business activities. More specifically entrepreneurial financial knowledge is assessed through five questions asking about knowledge of dividends, equity, mortgages, inflation, and risk-return relationship. Entrepreneurial financial behavior is assessed through nine questions asking about management of business bank account, business accounting, strategies to cope with theft, business data protection, business planning, business financing, and profitability. Finally, entrepreneurial financial attitude is assessed through three questions asking about long-term business financial goals, business external financing, and business financial planning. The overall score of financial literacy is represented by the sum of the scores obtained in each of the three components and thus ranges between 0 and 17.

These measurements too of course have their advantages and limitations. Certainly, considering the large amount of questions and the multiple components of financial literacy, these indicators appear as very accurate and exhaustive. However, at the same time, a very large amount of questions may increase the complexity of the measurement and make it difficult to replicate in other investigations. In addition, while for knowledge questions, there is only one answer that is definitely correct, the same does not apply to behavior and attitude questions. In fact, even if the questions on behaviors and attitudes consider one answer to be more correct than others, different opinions may emerge on which of the behaviors or attitudes are more or less correct. For these reasons, in this case too, there are examples of authors that modify these measurements and develop alternative ones, such as Lamla and Vinogradov (2019), who employ in their study a proxy of financial literacy composed of seven questions, where only four are taken from the OECD/INFE toolkit.

2.4 Measurement Error and Endogeneity Issues

As mentioned in the previous sections, measuring financial literacy is not an easy task. It is hard, in fact, to obtain an indicator without some limitations, and often some difficulties occur in these measurements. A typical issue related to financial literacy measurements and to the relationships between financial literacy and financial decisions and outcomes is represented by the measurement error. The issue of measurement error is quite relevant in the social sciences, and researchers often incur in difficulties in measuring concepts such as attitudes, behaviors, or abilities (Cernat 2021). Survey data are notoriously affected by measurement errors that may arise for many reasons (Bound et al. 2001) such as the level of effort or attention of the respondents, random answers, and financial literacy measures, as described in the previous sections, are usually based on questionnaire data. These empirical studies where financial literacy is computed as the sum of correct answers to a set of dedicated questions are called by Fernandes et al. (2014) as studies of "measured financial literacy", and these "measured financial literacy" studies are likely affected by measurement error.

Measurement error in these indicators, besides not capturing the real individual financial literacy, may also generate distortions, spill over to other variables, and underestimate or overestimate the effect of financial literacy on financial outcomes (Gillen et al. 2019; Lanciano et al. 2024). When studying the relationships between financial literacy and financial decisions and outcomes, measurement error, together with omitted variables and reverse causality, is a potential source of endogeneity and bias in the coefficients estimated with regressions (Wooldridge 2010).

The most common solution applied for solving endogeneity concerns in financial literacy studies is the Instrumental Variable (IV) approach (Fernandes et al. 2014). This method requires the use of an additional variable (the instrumental variable) that isolates the part of the independent variable that is uncorrelated with the error terms of the regression model. The choice of an instrumental variable requires a theoretical justification, and additionally, this variable has to be strongly correlated with the endogenous variable (financial literacy in this case) in order to be an instrument that cannot be considered as weak (Semadeni et al. 2014).

The existing literature on financial literacy reports several examples of studies employing instrumental variables. In most cases, these variables represent aspects related to the individual financial experience, such as learning experiences or family characteristics, or related to the level of financial literacy in a specific geographical area (Bucher-Koenen and Lusardi 2011; Behrman et al. 2012; Cucinelli and Soana 2023). Lusardi and Mitchell (2014) illustrate some of these examples of instrumental variables: Agnew et al. (2013), Alessie et al. (2011), Arrondel et al. (2012) and Fornero and Monticone (2011) instrument financial literacy by using financial experiences, information, and educational level of parents or siblings, assuming that financial and educational background of families may affect the individual level of financial literacy; Van Rooij et al. (2012) employ as instrumental variables for financial literacy the financial education that an individual acquired in school, while Christiansen et al. (2008) use the opening of universities, assuming that this induces high school graduate students in choosing an economics education; finally, the other examples reported are provided by Bucher-Koenen and Lusardi (2011), Sekita (2011) and Klapper et al. (2012) that use as instrumental variables respectively the exposure to financial knowledge of others in the same region, language abilities and the number of newspaper and universities. Further variables that may be used for instrumenting financial literacy are, for example, the number of Points of Sale (POS) and Automated Teller Machines (ATMs) in a determined geographical area, which is considered as an indicator of financial inclusion and thus as a potential driver for financial literacy (GPFI 2014), or numeracy skills of individuals or family members, which are considered as relevant drivers for financial literacy and aware financial decision making (Lusardi 2012).

As previously mentioned, another endogeneity problem that typically affects financial literacy studies is reverse causality. For instance, when testing whether financial literacy has an effect on a financial outcome, for example, participation to financial markets, the reverse relationship may also be possible, i.e., that participation to financial markets affects individual financial literacy. The issue is open to debate, but in general, in financial literacy studies using the IV approach, the results are generally confirmed, thus suggesting that, at least initially, it is financial literacy that has an impact on financial decisions (Lusardi and Mitchell 2014).

Finally, other solutions for dealing with the main limitations of financial literacy measurements can also consist of using multiple measurements, including alternative indicators or only some components of the original measurements. There are, indeed, many other ways for measuring financial literacy beyond the most employed indicators (Rieger 2020). These measurements can be adaptable according, for example, to the population studied or to specific goals of a financial education program or initiative. Alternative indicators, by allowing measurements to be adapted for specific purposes and defined concepts, are important for more accurately investigating the causality of specific relationships in studies with precise goals.

Examples of alternative indicators for financial literacy are provided, for instance, by French and McKillop (2016) and Sun et al. (2020). French and McKillop (2016) in their study reject to use the Big Three or Five questions proposed by Lusardi and Mitchell (2008, 2011), considered as not appropriate for the analyzed sample and for the goals of their research, while they employ an alternative financial literacy indicator which uses questions about numeracy abilities and money management skills, with the latter assessed through a nine-item scale developed by Garðarsdóttir and Dittmar (2012), which includes questions about personal finance management decisions and behaviors. Sun et al. (2020) while studying the effect of investors' financial literacy on corporate earnings quality, employed a specific proxy for assessing investors' financial literacy following Jappelli and Padula (2013), composed of questions about prices and percentages computation, together with questions about savings and interest rates. Otherwise, considering the three components in which OECD defines financial literacy, another idea could be to split this measure into its three components and to use some or one of them, such as only financial knowledge (Aristei and Gallo, 2021).

2.5 Critical Perspectives on Financial Literacy and Further Research Approaches

Despite the several efforts from experts, scholars, and international organizations, financial literacy theories and measurements are not without limitations and weaknesses. This section, in fact, introduces and discusses critical perspectives on the traditional research approaches related to financial literacy together with further research approaches proposed, available from the existing literature. These critical and alternative views are based on the assumption that other techniques and other disciplinary approaches are able to integrate or modify the existing ones, which are considered as mainstream.

According to this mainstream approach, usually, financial literacy studies are set up in order to verify through econometric models whether the level of financial literacy has an effect on a specific financial decision or outcome. These kinds of studies are the most frequent in the financial literacy literature. In these studies, identified as studies of "measured financial literacy" by Fernandes et al. (2014), usually financial literacy is measured through indicators which are composed by the percentage of correct answers to a certain number of questions. The use of these universal indicators, such as the Big Three, the Big Five questions or the OECD/INFE toolkit, causes some limitations. As already mentioned, the Big Three and Five could be indeed too synthetic and not flexible, while the OECD/INFE toolkit may be too complex and include debatable answers. Beyond these disadvantages, there are studies that show inconsistencies and incompatibilities between these financial literacy indicators. Van Hove and Ahunov (2024), for instance, show how the use of different financial literacy indicators causes different findings in the same relationships. These authors replicate two studies analyzing respectively the relationships between national culture and country-level financial literacy and the impact of financial literacy on financial inclusion. Both these studies employed originally the Standard & Poor's indicator, employed in the Global Financial Literacy Survey, which measures financial literacy as the percentage of correct answers to four questions asking about interest rates, interest compounding, inflation, and risk diversification. This indicator was substituted with the Big Three and OECD/INFE metrics. Using these alternative measurements, in the first study, many of the relationships between national culture and country-level financial literacy became weakly significant or totally disappeared, while in the second study, the magnitude of the coefficients on financial literacy in the regressions substantially decreased. A further study from Rieger (2020) analyzes the correlations between eight different scales of financial literacy proposed in literature, including the Big Three, finding that often correlation coefficients are relatively small and sometimes not statistically significant, and thus that different metrics often measure entirely different variables. These findings suggest that the results of financial literacy studies may change according to the metrics used.

The use of financial literacy measurements is indeed a very debated topic and different perspectives emerge in this discussion (Romagnoli and Trifilidis 2013; Montanaro and Romagnoli 2016; Kaiser et al. 2020). For example, considering the limitations of metrics aiming at assessing financial literacy through objective questions, based on simple tests, several studies also introduced subjective questions asking to individuals to self-assess their level of financial literacy. In fact, a positive relationship between subjective and objective financial literacy usually occurs (Allgood and Walstad 2013; Gignac 2005). The aim of using these questions is to capture the psychological factors that may influence an individual's financial decision-making process, and thus, they could be particularly useful in representing individual behaviors and attitudes in financial decision-making (Bellofatto et al. 2018). Surely, even for subjective measurements, there are evident limitations. While objective metrics, for instance, may have a limited number of questions or may include random answers or assess in the same way individuals responding correctly to the same amount of questions but with different levels of difficulty (D'Alessio et al. 2020), subjective metrics may overestimate the actual individual skills (Kruger and Dunning 1999). Therefore, it seems that, in general, individual skills and competencies, such as financial literacy, can only be measured with indicators that only partially capture the phenomenon.

Besides the so-called studies of "measured financial literacy", an alternative research approach includes studies where an experimental or quasi-experimental approaches are applied, that aim to assess the effects of financial education intervention on the level of financial literacy, which are defined by Fernandes et al. (2014) as studies of "manipulated financial literacy". These studies, less frequent in the financial literacy literature, are usually based on research conducted on the field, which includes the evaluation of a treatment effect, where the treatment is a financial education intervention. The experimental setting attempts to overcome specific limitations of the traditional multiple-choice questions survey designs previously explained (Muñoz-Murillo et al., 2020).

Among these studies, conflicting evidence emerges. There are examples of initiatives with positive effects, but financial education does not always work. Among the cases where a positive effect of financial education initiatives is reported, it is possible to mention studies from Kaiser and Menkhoff (2020), Carpena and Zia (2020), Amagir et al. (2018),

Becchetti et al. (2013), Boyce and Danes (1998), Mandell (2008), or Bernheim et al. (2001), which show how financial education courses or initiatives, often delivered among school students, positively affect the individual level of financial literacy among different contexts and targets.

But, as mentioned, sometimes the effects of financial education interventions are surprisingly weak (Gneezy et al., 2011), thus generating a pessimistic skepticism on financial education and literacy among some authors. Fernandes et al. (2014), by conducting a meta-analysis of 168 studies investigating the relationships between financial education, financial literacy, and financial behaviors, divided these studies in studies of "measured financial literacy" and "manipulated financial literacy". In their results, they find an average positive effect of financial literacy on financial behaviors in the "measured financial literacy studies", while they find an average very slight effect of financial education interventions aiming at improving financial literacy. Thus, they find significantly smaller effects in the experimental studies. The experimental approach clearly has its limitations too. For instance, the effect of the interventions may decay over time, thus having an effect only for a limited time period, or otherwise, there may be weaknesses in the design of the experiments or the interventions that could affect their effectiveness.

Therefore, the comparison between different metrics and research approaches shows how often the methodology employed may affect the effects of financial literacy or education and thus the findings of the studies.

Finally, it is possible to mention further research approaches that include other disciplines. Indeed, according to various authors, financial literacy and education are topics that shouldn't be associated only with economics or finance experts. Authors and scholars from other disciplines, such as sociology, psychology, or pedagogy, also deal with financial literacy and education and employ their typical research approaches by including sociological, psychological, or pedagogical theories and perspectives in financial literacy and education studies. Examples are provided by Rinaldi (2007), Rinaldi and Giromini (2002), or Berti et al. (1986), which studied the determinants of knowledge and understanding of concepts related to money such as price, payment, cost, or profit among children, or how children acquire specific attitudes towards money, and consider, in their research design, theories explaining the stages of awareness development about the concept of money and its value (Berti & Bombi 1981) such as the theories of the economic or consumer socialization, which is the process by which young people or consumers acquire skills, knowledge, and attitudes related to money, relevant for living in the economic world (Ward 1974; Denhardt and Jeffress 1971; Webley et al. 2004). Considering these theories could be useful indeed for selecting the right topics for financial education initiatives and interventions directed to young people based on their understanding and learning abilities.

2.6 Conclusions

This chapter has shown and explained that financial literacy is a concept that can be measured in different ways, from the Big Three or Five questions, to the OECD/INFE toolkit, to more personalized measures or measures defined according to specific goals of the studies.

Each of these measures has its advantages and disadvantages, and it is important that those who use them know how to exploit their positive aspects but also take into account the limits of these measures, to guarantee the maximum possible reliability of the evidence that emerges from scientific studies.

Over the years, various schools of thought have also developed, which approach the topic of financial literacy in different ways and in some cases have divergent opinions. Concerning this last aspect, the content of this chapter also represents a good opportunity to learn about these different schools of thought and the main authors that populate the literature scene.

Therefore, the overall goal of this chapter is to provide the reader with a basic knowledge of measurement techniques, an awareness of the availability of multiple solutions, a general understanding of the world of the research on the topic of financial literacy, and more specifically, of the different approaches that are employed.

References

Agnew, J., Bateman, H., Thorp, S. (2013). Financial Literacy and Retirement Planning in Australia. Numeracy 6 (2).

Alessie, R., Van Rooij, M., Lusardi, A. (2011). Financial Literacy and Retirement Preparation in the Netherlands. Journal of Pension Economics and Finance 10 (4): 527–45.

Allgood, S., Walstad, W. (2013). "Financial Literacy and Credit Card Behaviors: A Cross-Sectional Analysis by Age." Numeracy 6 (2).

Amagir, A., Groot, W., Van Den Brink, H., Wilschut, A. (2018). A review of financial literacy education programs for children and adolescents. Citizenship, Social and Economics Education 2018, Vol. 17(1) 56–80.

Anderson, A., Robinson, D. T. (2022). Financial literacy in the age of green investment. Review of Finance, 26(6), 1551–1584.

Angrisani, M., Burke, J., Lusardi, A., Mottola, G. (2023). The evolution of financial literacy over time and its predictive power for financial outcomes: evidence from longitudinal data. Journal of Pension Economics and Finance, 22(4), 640-657.

Aristei, D., Gallo, M. (2021). Financial Knowledge, Confidence, and Sustainable Financial Behavior. Sustainability, 13(19), 10926.

Arrondel, L., Debbich, M., Savignac, F. (2012). Stockholding and Financial Literacy in the French Population. International Journal of Social Sciences and Humanity Studies 4 (2).

Becchetti L., Caiazza S., Coviello D. (2013), «Financial Education and Investment Attitudes in High Schools: Evidence From a Randomized Experiment», in Applied Financial Economics, 23:10, pp. 817-836.

Behrman, J. R., Mitchell, O. S., Soo, C. K., Bravo, D. (2012). How financial literacy affects household wealth accumulation. American Economic Review, 102(3), 300-304.

Bellofatto, A., D'Hondt, C., De Winne R. (2018). Subjective financial literacy and retail investors' behavior. Journal of Banking and Finance 92 (2018) 168–181.

Bernheim, B. D., Garrett, D. M., Maki, D. M. (2001) Education and saving: the long-term effects of high school financial curriculum mandates, Journal of Public Economics, 80, 435–65.

Berti, A. E., & Bombi, A. S. (1981). The Development of the Concept of Money and Its Value: A Longitudinal Study. Child Development, 52(4), 1179–1182.

Berti, A. E., Bombi, A. S., & De Beni, R. (1986). Acquiring economic notions: Profit. International Journal of Behavioral Development, 9(1), 15–29.

Bound, J., C. Brown, and N. Mathiowetz (2001). Chapter 59- measurement error in survey data. Volume 5 of Handbook of Econometrics, pp. 3705–3843. Elsevier.

Boyce, L. and Danes, S. (1998) Evaluation of the NEFE High School Financial Planning Program, National Endowment for Financial Education (NEFE), Engelwood, CO.

Bucher-Koenen, T., Lusardi, A. (2011). Financial literacy and retirement planning in Germany. Journal of Pension Economics & Finance, 10(4), 565-584.

Carpena, F., Zia, B. (2020). The causal mechanism of financial education: Evidence from mediation analysis. Journal of Economic Behavior and Organization 177 (2020) 143–184.

Cernat, A. (2021). Measurement Error in Longitudinal Data. Oxford University Press.

Christiansen, C., Schroter Joensen, J., Rangvid, J. (2008). Are Economists More Likely to Hold Stocks? Review of Finance 12 (3): 465–96.

Cucinelli, D., Soana, M. (2023). Investor preferences, financial literacy and intermediary choice towards sustainability. Research in International Business and Finance, 66, 102027.

D'Alessio, G., de Bonis, R., Neri, A., Rampazzi, C. (2020). Financial literacy in Italy: the results of the Bank of Italy's 2020 survey, Bank of Italy.

Denhardt, R. B., & Jeffress, P. W. (1971). Social Learning and Economic Behavior: The Process of Economic Socialization. The American Journal of Economics and Sociology, 30(2), 113–125.

Fernandes, D., J. G. Lynch Jr, and R. G. Netemeyer (2014). Financial literacy, financial education, and downstream financial behaviors. Management Science 60(8), 1861–1883.

Fornero, E., Monticone, C. (2011). Financial Literacy and Pension Plan Participation in Italy. Journal of Pension Economics and Finance 10 (4): 547–64.

French, D., McKillop, D. (2016). Financial literacy and over-indebtedness in low-income households. International Review of Financial Analysis, 48 (2016) 1–11.

Garðarsdóttir, R., Dittmar, H. (2012). The relationship of materialism to debt and financial well-being: The case of Iceland's perceived prosperity. Journal of Economic Psychology, 33(3), 471–481.

Gignac, G. (2005). The association between objective and subjective financial literacy: Failure to observe the Dunning-Kruger effect. Personality and Individual Differences. 184 (2022), 111224.

Gillen, B., Snowberg, B., Yariv, L. (2019). Experimenting with measurement error: Techniques with applications to the caltech cohort study. Journal of Political Economy, vol 127(4), pages 1826-1863.

Gneezy U, Meier S, Rey-Biel P (2011) When and why incentives (don't) work to modify behavior. Journal of Economic Perspectives. 25(4): 191-210.

GPFI, Global Partnership for Financial Inclusion (2014). G20 financial inclusion indicators.

Hastings, J.S., Madrian, B.C., Skimmyhorn, W. (2013). Financial literacy, financial education and economic outcomes. Annual Review of Economics 5, 347–373.

Jappelli, T., Padula, M. (2013). Investment in financial literacy and saving decisions, Journal of Banking & Finance, Volume 37, Issue 8, 2779-2792.

Kaiser T., Menkhoff L. (2020), Financial education in schools: A meta-analysis of experimental studies, Economics of education review (78).

Kaiser, T, Lusardi, A., Menkhoff, L., Urban, C. J. (2020), Financial education affects financial knowledge and downstream behavior, Nber Working Paper Series, n. 27057.

Klapper, L., Lusardi, A., Panos, G. (2012). Financial Literacy and the Financial Crisis. National Bureau of Economic Research Working Paper 17930.

Kruger, J., Dunning, D. (1999). Unskilled and unaware of it: How difficulties in recognizing one's own incompetence lead to inflated self-assessments. Journal of Personality and Social Psychology, 77(6), 1121–1134.

Lamla, M. J., & Vinogradov, D. V. (2019). Central bank announcements: Big news for little people? Journal of Monetary Economics, 108, 21–38.

Lanciano, E., Previati, D., Ricci, O., Stentella Lopes, S. (2024). Gender differences and measurement error in financial literacy. Journal of Behavioral and Experimental Finance, Volume 41, 2024, 100896.

Lusardi, A. (2012). Numeracy, financial literacy and financial decision making. Numeracy, 5(1), Article 2, January 2012.

Lusardi, A., & Mitchell, O. S. (2008). Planning and financial literacy: How do women fare?. American Economic Review, 98(2), 413-417.

Lusardi, A., Mitchell, O. (2011). Financial literacy around the world: an overview. Journal of Pension Economics and Finance, 10, pp 497-508,

Lusardi, A., Mitchell, O. (2014). The economic importance of financial literacy: Theory and evidence. Journal of Economic Literature, 52(1), 5–44.

Mandell, L. (2008) High school financial literacy, in Overcoming the Savings Slump. How to Increase the Effectiveness of Financial education: a randomized experiment 835 Financial Education and Savings Programs (Ed.) A. Lusardi, University of Chicago Press Chicago, IL, pp. 257–79.

Montanaro, P., A. Romagnoli (2016). Financial literacy in PISA 2012: an analysis of results and the role of families in Italy, Bank of Italy, Questioni di Economia e Finanza (Occasional Papers), n. 335.

Muñoz-Murillo, M., Álvarez-Franco, P., Restrepo-Tobón, D. (2020). The role of cognitive abilities on financial literacy: New experimental evidence, Journal of Behavioral and Experimental Economics, Volume 84, 101482.

OECD (2020). OECD/INFE survey instrument to measure the financial literacy of MSMEs.

OECD/INFE (2022) Toolkit for Measuring Financial Literacy and Financial Inclusion.

Ranyard, R., McNair, S., Nicolini, G., & Duxbury, D. (2020). An item response theory approach to constructing and evaluating brief and in-depth financial literacy scales. Journal of Consumer Affairs, 54(3), 1121–1156.

Rieger, M. (2020). How to measure financial literacy. Journal of Risk and Financial Management, 13, 324.

Rinaldi E. (2007), Giovani e denaro. Percorsi di socializzazione economica, Unicopli

Rinaldi, E., and Giromini, E. (2002). The importance of money to Italian children. International Journal of Advertising and Marketing to Children, July-September, pp. 53–59

Romagnoli, A., M. Trifilidis, (2013). Does financial education at school work? Evidence from Italy, Bank of Italy, Questioni di Economia e Finanza (Occasional Papers), n. 155.

Sekita, S. (2011). Financial Literacy and Retire ment Planning in Japan. Journal of Pension Economics and Finance 10 (4): 637–56.

Semadeni, M., Withers, M., Trevis Certo, S. (2014). The perils of endogeneity and instrumental variables in strategy research: Understanding through simulations. Strategic Management Journal 35(7), 1070–1079.

Sun, H., Yuen, D., Zhangc, J., Zhangb, X. (2020). Is knowledge powerful? Evidence from financial education and earnings quality. Research in International Business and Finance 52(2020) 101179.

Van Hove, L., & Ahunov, M. (2024). Financial literacy: different indicator, different insights? Cogent Economics & Finance, 12(1).

Van Rooij, M., Lusardi, A., & Alessie, R. (2011). Financial literacy and stock market participation. Journal of Financial Economics, 101(2), 449–472.

Van Rooij, M., Lusardi, A., Alessie, R. (2012). Financial Literacy, Retirement Planning and Household Wealth. Economic Journal 122 (560): 449–78.

Ward, S. (1974). Consumer Socialization. Journal of Consumer Research, 1(2), 1 14. https://www.jstor.org/stable/2489100

Webley, P., Burgoyne, C.B., Lea, S.E.G., Young, B.M. (2002). The Economic Psychology of Everyday Life. Psychology Press.

Wooldridge, J. M. (2010). Econometric analysis of cross section and panel data. MIT press.

CHAPTER 3

Financial Literacy and Related Financial Outcomes

Abstract Why is financial literacy important? What is the relevance of financial literacy for financial decisions and outcomes? These are the main questions that this chapter attempts to answer, through evidence from the main high-impact scientific studies available in the literature on financial literacy. Specifically, this chapter investigates the relationship between financial literacy and various specific financial decisions and outcomes, such as saving, investments, retirement, or insurance, through the collection of findings from the literature, which describes these relationships in different contexts and among various targets, showing how financial literacy may positively affect savvy financial behaviors and the overall economic well-being of individuals and society.

Keywords Financial literacy · Financial behavior · Financial decisions · Financial outcomes

E. Lanciano, *The Relevance of Financial Literacy*,
https://doi.org/10.1007/978-3-031-93397-4_3

3.1 Introduction

This chapter addresses a central aspect of the book, i.e., the concrete relevance of financial literacy in various situations. Despite the skepticism of some authors and the development of some critical views, the literature seems to generally agree that financial literacy has a tangible effect on financial decisions and outcomes.

By collecting and discussing the results available in the literature on multiple financial literacy studies, the chapter describes how, for example, greater financial literacy corresponds to a higher propensity to save or to a lower financial vulnerability, making individuals and businesses better prepared to face difficult times.

The chapter also describes how financial literacy improves the chances of an individual or an entrepreneur for obtaining credit, or better credit conditions, and therefore greater trust from financial intermediaries, or how it leads to more informed and prudent investment decisions and a greater likelihood of obtaining better returns.

Finally, the chapter focuses on the positive effect that financial literacy may have on thinking adequately about how to financially plan for retirement, and on the ability to evaluate and develop strategies to protect against risks through insurance products.

The relationship between financial literacy and these financial needs is also evident in the questions used in financial literacy measurements, which often ask about knowledge of these concepts or whether appropriate behaviors are applied in these areas.

The fact that these findings come from multiple contexts and countries confirms their international validity and the consistency of the results, despite the use of various measures and different samples.

3.2 Savings and Financial Resilience

Despite the different views and schools of thought that have developed over the years on financial literacy, it is possible to state that, generally, the research world has concluded that financial literacy is associated with various healthy financial behaviors (Fernandes et al. 2014). According to Lusardi and Mitchell (2014), financial literacy provides necessary skills for taking aware and possibly correct financial decisions. The relevance of financial literacy has additionally increased over the years considering the frequent economic and financial crises of the last 15 years, in which people were called to take cautious financial decisions and manage difficult situations. Financial literacy is indeed associated with a general better resource management, long-term economic planning, higher propensity to save money, more preparation to unpredicted events, higher awareness of investment decisions, better liability management, retirement planning, and risk protection strategies such as signing an insurance contract, while lower financial literacy usually causes inconvenient situations such as more expensive loans, incurring in higher transaction costs, having excessive debt exposures, inability to accumulating savings, or more time spent worrying about personal finances.

The relevance of financial literacy does not refer only to individuals but affects entrepreneurs as well, especially owners of micro or small enterprises. Various studies report how entrepreneurial financial literacy improves management practices such as budgeting, credit management, building liquidity buffers or improves the profitability and growth of enterprises (D'Ignazio et al. 2023; Dahmen and Rodríguez 2014; Eniola and Entebang 2017; Atandi 2021; Finaldi Russo et al. 2022; Babajide et al. 2023).

In this section, the relevance of financial literacy for savings decisions and financial resilience is discussed. According to the literature, in fact, financially literate people are more likely to accumulate savings (De Bassa Scheresberg 2013) for increasing their financial resilience (Clark et al. 2021) and reducing their financial vulnerability. The positive relationship between financial literacy and saving principles is persistent among experimental and non-experimental studies conducted in several contexts including western and non-western countries (Fernandes et al. 2014; Estrada-Mejia et al. 2023). This relationship is analyzed with different perspectives, related to the specific goals of saving decisions. The ability to save is indeed related to various goals. Saving could be useful for accumulating wealth for investing, for retirement planning reasons, or for creating an emergency fund for precautionary reasons and for dealing with specific vulnerabilities (Lusardi and Mitchell 2007).

About emergency situations, the relevance of savings and financial resilience has significantly increased during and after the COVID pandemic. In that period, the literature that analyzes the relationship between financial literacy and savings or financial resilience was particularly fertile, and the findings of these studies generally show how financially literate people had a stronger resistance to the pandemic shock thanks to their precautionary savings, being more prepared to use liquidity buffers for facing unpredicted events, or for coping with periods of lack of income (Klapper and Lusardi 2020). These findings thus show how financial literacy should be not considered as important only in ordinary times but especially in extraordinary circumstances. A typical question used in financial literacy investigations for assessing the ability to save asks how much an individual is confident to come up with 2000 euro if an unforeseen event occurs. Usually, financially literate people are more confident, while people with lower literacy show higher financial fragility (Hasler and Lusardi 2017).

The relationship between financial literacy and financial resilience is also evident from the questions used in financial literacy measurements, aiming at investigating savings behaviors. Saving outcomes are thus used for assessing the individual level of financial literacy. In particular, in the OECD/INFE toolkit, in the sets of questions dedicated to financial behaviors and attitudes, there are specific questions asking whether individuals actively save or whether they prefer to spend or to save money for long-term goals. Thus, it is quite clear that financial literacy is considered as a relevant driver for savings, financial resilience, and wealth accumulation for increasing individual and social economic well-being (Demertzis et al. 2020; Clark et al. 2021; Hurwitz et al. 2021).

3.3 Access to Credit

Financial literacy also seems to have a relevant effect on access to credit. Among the studies that have identified positive relationships between financial literacy and access to credit, it is possible to mention Lusardi and Mitchell (2014) who found that, among a variety of economic indicators, financial literacy also affects using bank credit. Moreover, financial literacy does not affect only access to credit but also obtaining better credit conditions. Fernandes et al. (2014), in their meta-analysis of the relationships between financial literacy and financial behaviors and outcomes, among the analyzed studies found an overall positive relationship between financial literacy and credit scores, and a negative relationship between financial literacy and the amount of credit fees. The positive relationship between financial literacy and higher credit scores is also highlighted by Frisancho (2023).

Financial illiteracy, on the other hand, is often associated with higher cost loans, excessive debt exposures, and higher probabilities of credit default. The link between financial literacy and access to formal credit is persistent among several contexts, emerging also in studies conducted in Southern American countries such as Peru and Uruguay (Estrada-Mejìa et al. 2023; Sotomayor et al. 2018; Roa et al. 2019; Santos et al. 2018).

This effect of financial literacy on access to credit has to be considered not only for individuals but also among entrepreneurs. Entrepreneurial financial literacy has indeed a relevant role for raising credit, especially in enterprises of small dimensions. Financial literacy of small and micro business owners, beyond providing higher skills in debt management and financial planning, may also improve the individual relationships between the entrepreneurs and financial intermediaries, thus being an overall potential relevant driver for obtaining credit. An example of the effect of financial literacy on access to credit in enterprises of small dimensions is provided by Calcagno et al. (2024), which analyzed this relationship on a sample composed of almost 2000 micro, small, or medium-sized enterprises. This study was conducted in Italy, a country where micro and small businesses represent almost all the enterprises in the country. The research, in cooperation with Bank of Italy, shows how financially literate entrepreneurs are more likely to obtaining credit from banks.

Credit knowledge and credit management abilities are also aspects considered in the assessment of the overall level of individual and entrepreneurial financial literacy. The OECD/INFE toolkits for measuring financial literacy of individuals and entrepreneurs include questions aiming at investigating the basic understanding of credit and debt management abilities, by asking simple numerical calculations on the interest on a loan (You lend $25 to a friend one evening and he gives you $25 back the next day. How much interest has he paid on this loan?), knowledge of the relationship between the interest paid and the life of a loan (A 15-year loan typically requires higher monthly payments that a 30-year loan, but the total interest paid over the life of the loan will be less), and finally by asking whether an individual or an entrepreneur pays regularly their bills and liabilities.

Thus, financial literacy seems to generally have a positive role in obtaining credit. This positive influence can be explained in various ways. For example, financial literacy, by increasing an individual's awareness and confidence with financial matters, can reduce the discouragement that typically affects many individuals and entrepreneurs in applying for credit because they are afraid of being turned down or because the application costs are too high (Kon and Storey 2003). Or, an individual or an entrepreneur with higher debt management skills improves credit conditions, reduces banks' screening and monitoring costs, and banks' expected losses (Wise 2013; Hussain et al. 2018; Calcagno et al. 2024).

3.4 Investment Behavior and Decisions

Financial literacy affects also investment decisions and positive investment behaviors. This relationship emerges immediately from some questions used for financial literacy measurements, such as the Big Three or Five questions. More specifically, knowledge and understanding of basic investment principles, such as risk diversification and risk-return relationship, are investigated for assessing the individual level of financial literacy. An adequate level of financial literacy, indeed, requires that an individual knows that the investment portfolio should be diversified in order to mitigate the overall risk of the investments, and that potential higher investment returns correspond to higher investment risks. Knowledge of the concept of risk diversification is investigated through a question asking whether the respondent agrees or not that buying a single company stock usually provides a safer return than a stock mutual fund, while the understanding of the relationship between investment risks and returns is investigated through a question about bond prices and interest rates, asking to the respondent what happens to bond prices if the interest rates rise.

Investment understanding and decisions are also investigated in the OECD/INFE toolkits for measuring financial literacy, which include questions on risk diversification, risk-return relationships, and investment decisions. The financial knowledge section of these toolkits contains questions asking whether it is possible to reduce the risk of investing in the stock market by buying a wide range of stocks and shares, or whether an investment with a high return is likely to have high risk, or, for entrepreneurs, whether they know what dividends and equity are. Additionally, investment decisions are investigated in the financial behavior section of these toolkits, through questions asking to the respondents whether they buy bonds, invest in stocks, invest in crypto-assets or in some other ways.

Financial literacy is considered also as a driver for developing more awareness and avoiding the so-called behavioral traps that may typically affect investment decisions such as overconfidence, risk propensity, framing effect, disposition effect, or loss aversion. Behavioral traps are phenomena that are typically studied in the field of behavioral finance. These traps may indeed lead the investor into wrong decisions due to emotional reactions or incorrect evaluations. Overconfidence, usually measured through a question asking individuals to self-assess from 1 to 10 their level of financial skills, causes an overestimation of individual capabilities that may induce to take not completely aware decisions (Chhatwani and Parija, 2023). The same is for risk propensity, which is usually assessed through questions asking to individuals how much they feel confident to take risks in financial decisions on a 1–10 scale. Financially literate people are usually less likely to take unconscious risks (Fernandes et al. 2014). Other examples of traps that may affect investment outcomes are represented by the loss aversion, which is an emotional reaction to losses stronger than the reaction to returns of the same amount, given by the greater sensitivity to the possibility of losing than to the possibility of gaining a given amount, or by the disposition effect, which is the attitude of investors to maintain assets in their portfolio that are in loss and, at the same time, to quickly sell assets with positive trends with the goal of securing a short term return, or by the framing effect, which is the way in which the information is represented, which can distort the investor's attention to information considered as relevant and modify his or her attitude towards risks. Thus, financial literacy is generally associated with more aware and prudent investment decisions and behaviors.

The main findings available from the literature on this relationship show how generally financial literacy increases the participation to financial markets and allows to take more informed investment decisions and to have more possibilities to identify better-performing assets or to pay lower investment fees. Lusardi and Mitchell (2023) found how individual financial literacy produces better investment outcomes, while Van Rooij et al. (2011) and Clark et al. (2017) show how individuals with higher financial literacy are more likely to invest in the stock market, and hence, earn higher returns on their investments. Bellofatto et al. (2018), in their study analyzing the relationships between individual financial literacy and investment behaviors and outcomes, found how investors with higher levels of financial literacy seem to invest smarter, trade more, are less prone to the disposition effect, own diversified portfolios, and report higher gross and net returns. A further positive effect of financial literacy on investment behaviors is the use of professional financial advisory, which allows to listen opinions and recommendations from an expert before taking an investment decision. The use of financial advisory is, in fact, often associated with higher propensity to invest (Gentile et al. 2015).

3.5 Retirement Planning

The time of retirement is an often-underestimated issue on the financial point of view. It is usually seen as a very distant time that is not worth thinking about in the present and so decisions are postponed to future times. However, it often happens that people come to think about it very close to this time, when it is too late to make financial decisions with long-term results such as savings or investments that can integrate the pension received from the public pension system. Moreover, it is not considered that nowadays the period of retirement is a very long time considering the average increase in life expectancy over the years, and thus necessitates an adequate financial planning to live this phase of life comfortably.

Financially literate people usually pay more attention to this issue and are more likely to plan carefully for their retirement through adequate saving or investment decisions taken at the right time.

As well as for the other outcomes analyzed in this chapter, retirement planning is also an investigated issue considered in the overall assessment of the individual level of financial literacy. The OECD/INFE toolkits for individuals and entrepreneurs contain specific questions asking to the respondents whether they set or not long-term financial goals, including retirement planning goals, or whether they think about how to fund their own retirement. Surveys on financial literacy usually contain other typical questions about retirement planning. For example, they include questions about knowledge of longevity risk, asking about the relationship between life expectancy and the amount of pensions, or questions asking to the respondents when they think they should start saving to integrate their public pension.

According to Lusardi and Mitchell (2007, 2011), financial literacy is a strong predictor of retirement planning and wealth accumulation. These findings are persistent across several contexts and countries (Behrman et al. 2012; Hastings and Mitchell 2020; Lusardi and Mitchell 2014). Examples of studies that replicated the analysis of this relationship are provided by Bucher-Koenen and Lusardi (2011), which show the positive impact of financial knowledge on retirement planning in Germany, or by Agnew et al. (2013), which assess the effect of financial literacy on retirement planning in Australia, or by Van Rooij et al. (2012), which found how financial literacy is positively related to retirement planning and the development of a savings plans for wealth accumulation in the Netherlands, or also by Estrada-Mejia C. et al. (2023) which found that financially literate people are more likely to be able to save and plan for retirement in Latin America. Positive effects of financial literacy on retirement planning and pension plan participation are reported also in Italy (Fornero and Monticone, 2011; Ricci and Caratelli, 2017), Sweden (Almenberg and Säve-Söderbergh, 2011) or Japan (Sekita 2011). Another interesting finding is provided by Gustman et al. (2010) who noted how financial knowledge is associated with greater knowledge of retirement plan rules.

Thus, several authors have linked financial literacy to retirement planning, generally finding how the more financially savvy people are more likely to save or invest for their retirement, and how a significant part of wealth inequalities in retirement is explained by different levels of financial literacy (Lusardi et al. 2017). The key role of financial literacy in financial preparation of retirement is probably one of the most robust findings across the literature on financial literacy (Adams and Rau 2011).

3.6 Insurance

The last financial outcome analyzed in this chapter is the insurance against risks. In fact, financial literacy is also positively associated with protection against risks through insurance contracts. Financially literate people are more likely to better assess the risks related to their goods, to their business, or to their life, to estimate their financial impact, and to transfer them to insurance companies for their covering.

Compared to the previous outcomes and decisions analyzed, the relationship between financial literacy and insurance product subscription is still quite underexplored (Bongini et al. 2023). However, despite this lack of studies, evidence of the positive effect of financial literacy on insurance is not missing. The role of financial literacy for risk management and protection strategies represented by insurance is highlighted by Lusardi and Mitchell (2014). Pitthan and De Witte (2021), through a comprehensive review of the literature on the studies dealing with these topics, explain in their study how financial literacy could mitigate the effects of behavioral biases and heuristics in insurance demand. Other positive findings are reported by Cole et al. (2013), Wang et al. (2021), and Cai et al. (2013), with the first finding how financial literacy was particularly beneficial for improving insurance demand in communities suffering from underinsurance in India, and the other two both finding positive associations between financial literacy and insurance demand in China.

Ricci and Santilli (2024) found significant evidence on this relationship in Italy. Specifically, they analyzed the relationship between entrepreneurs' financial literacy and the adoption of insurance for business interruptions due to the effect of the COVID pandemic, finding a positive association, and thus highlighting the key role of financial literacy in shaping risk management strategies for businesses and showing the increasing relevance of insurance products in covering risks related to similar events, such as natural catastrophes, that may have a negative impact on the business operations, supply chains or results. Other findings from Italy come from Bongini et al. (2023), who report in their study how insurance literacy, which is a specific component of financial literacy focused on insurance, has a positive relationship with both life and non-life insurance product purchases.

The adoption of insurance contracts is an aspect that is also investigated in financial literacy measurements. The OECD/INFE toolkit for entrepreneurs contains a precise question asking about strategies to cope with theft of the business equipment needed to operate, such as computers or vehicles. Specifically, this question asks whether the entrepreneur has covered or not this risk through a dedicated insurance.

Various other studies focus on the impact of financial literacy on insurance decisions, finding how individuals with higher financial literacy tend more to hold insurance products (Armantier et al. 2023; Lin et al. 2017; Allgood and Walstad 2016; Luciano et al. 2016). Thus, the overall evidence on financial literacy and insurance shows how financial literacy is relevant in insurance purchase decisions, for developing the insurance sector, and contrasting underinsurance. In general, greater literacy is associated with greater participation in the insurance market (Bongini et al. 2023).

The evidence gathered in this chapter represents part of the product of about 20 years of financial literacy studies. The studies analyzed have had a significant impact on the scientific community, considering also the level of publications. Moreover, although these studies have been conducted in different contexts and countries, are carried out on different samples and databases, and use different measures, they lead to similar and consistent results among themselves.

In order to provide a quick consultation of the literature related to specific topics, the following table (Table 3.1) represents the main sources available from the existing literature, mentioned in this chapter, where the relationships between financial literacy and these financial outcomes are analyzed.

3.7 Conclusions

This chapter shows how and why financial literacy is relevant, according to the findings of studies conducted by leading experts in the literature on the topic. The relevance of financial literacy in different financial areas and across different contexts highlights the transversal importance of the topic, which represents a key aspect in taking financial decisions related to the everyday life, not only for those who work in the field of economics and finance but for everyone.

The aim of this chapter is to make the reader understand precisely this, i.e., that financial literacy is relevant, plays a key role in financial decisions and outcomes, and therefore has a concrete impact on the everyday life, and that this is widely documented by several scientific studies.

The final table that summarizes the bibliographic sources analyzed for each relationship between financial literacy and financial outcomes discussed in this chapter can be a useful tool for the reader to have a defined picture of some of the studies that belong to certain fields and that analyze certain relationships, and of the main authors of reference, and that can be easily consulted.

Table 3.1 Bibliographic sources on financial literacy and financial outcomes

Financial outcomes	*Sources*
Savings and financial resilience	Atandi, F.G., (2021) "Role of entrepreneur's Competence on Growth of Small and Medium Enterprises", International Journal of Management & Entrepreneurship Research, 3 (2): 84–96 Babajide, A., Osabuohien, E., Tunji-Olayeni, P., Falola, H., Amodu, L., Olokoyo, F., Adegboye, F., Ehikioya B. (2023) Financial literacy, financial capabilities, and sustainable business model practice among small business owners in Nigeria. Journal of Sustainable Finance & Investment, 13:4, 1670–1692 Clark, R.L., Lusardi, A., Mitchell, O., (2021). Financial fragility during the COVID-19 pandemic. AEA Papers and Proceedings, 111, 292–296 D'Ignazio, A., Finaldi Russo, P., Stacchini, M., (2022). Micro-entrepreneurs' financial and digital competences during the pandemic in Italy, Bank of Italy Occasional Papers No. 724, October 2022 Dahmen, P. and E. Rodríguez, (2014) "Financial Literacy and the Success of Small Businesses: An Observation from a Small Business Development Center", Numeracy, 7(1) De Bassa Scheresberg, C. (2013). Financial Literacy and Financial Behavior among Young Adults: Evidence and Implications. Numeracy 6 (2) Demertzis, M., Domínguez-Jiménez, M., Lusardi, A. (2020). The Financial Fragility of European Households in the Time of COVID-19. Bruegel Policy Contribution 2020/15 Eniola, A. and H. Entebang, (2017) "SME Managers and Financial Literacy", Global Business Review, 18 (3): 559–576

(continued)

Table 3.1 (continued)

Financial outcomes	*Sources*
	Estrada-Mejia C., Diana Mejía, D., Cordoba, P. (2023). Financial literacy and financial wellbeing: Evidence from Peru and Uruguay. Journal of Financial Literacy and Wellbeing (2023), 1, 403–429 Fernandes, D., J. G. Lynch Jr, and R. G. Netemeyer (2014). Financial literacy, financial education, and downstream financial behaviors. Management Science 60(8), 1861–1883 Finaldi Russo, P., Galotto L. and C. Rampazzi (2022) "The financial literacy of microentrepreneurs: evidence from Italy", Bank of Italy Occasional Papers No. 727, October 2022 Hasler, A. Lusardi, A. (2017). The gender gap in financial literacy: A global perspective. Global Financial Literacy Excellence Center, The George Washington University School of Business Hurwitz, A., Mitchell, O., Sade, O. (2021). Longevity Perceptions and Saving Decisions during the COVID-19 Outbreak: An Experimental Investigation. AEA Papers and Proceedings, 111: 297–301 Klapper, L., Lusardi, A. (2020). Financial literacy and financial resilience: Evidence from around the world. Financial Management 49(3), 589–614 Lusardi, A. & Mitchell, O. S. (2007). Baby boomer retirement security: The roles of planning, financial literacy, and housing wealth. Journal of Monetary Economics, 54, 205–224 Lusardi, A., Mitchell, O. (2014). The economic importance of financial literacy: Theory and evidence. Journal of Economic Literature, 52(1), 5–44

(continued)

Table 3.1 (continued)

Financial outcomes	*Sources*
Access to credit	Calcagno, R., Finaldi Russo, P., Galotto, L., Quas, A. (2024). Financial literacy of micro entrepreneurs and access to credit. Bank of Italy, Occasional papers, number 853, June 2024 Estrada-Mejia C., Diana Mejía, D., Cordoba, P. (2023). Financial literacy and financial wellbeing: Evidence from Peru and Uruguay. Journal of Financial Literacy and Wellbeing (2023), 1, 403–429 Fernandes, D., J. G. Lynch Jr, and R. G. Netemeyer (2014). Financial literacy, financial education, and downstream financial behaviors. Management Science 60(8), 1861–1883 Frisancho, (2023). Spillover Effects of Financial Education: The Impact of School-Based Programs on Parents. Journal of Financial Literacy and Wellbeing 1: 138–153 Hussain, J., Salia, S., Karim, A. (2018). Is knowledge that powerful? Financial literacy and access to finance: an analysis of enterprises in the UK. Journal of Small Business Entrepreneurial Development, 25(6):985–1003 Lusardi, A., Mitchell, O. (2014). The economic importance of financial literacy: Theory and evidence. Journal of Economic Literature, 52(1), 5–44 Kon, Y., Storey, D.J. (2003). A theory of discouraged borrowers. Small Business Economics 21, 37 49 Roa, M., Garron, I., Barboza, J. (2019). Financial Decisions and Financial Capabilities in the Andean Region. Journal of Consumer Affairs 53: 296–323 Santos, D., Da Silva, W., Gonzalez, L. (2018). Lower Financial Literacy Induces Use of Informal Loans. RAE– Revista de Administração de Empresas 58: 44–59 Sotomayor, N., Talledo, J., Wong, S. (2018). Determinants of financial inclusion in Peru: Recent Evidence from the Demand Side. SBS Working Paper No. DT/06/2018 Wise, S. (2013). The Impact of Financial Literacy on New Venture Survival", International Journal of Business and Management; 8 (23): 30–39

(continued)

Table 3.1 (continued)

Financial outcomes	*Sources*
Investment behaviors and decisions	Bellofatto, A., D'Hondt, C., De Winne R. (2018). Subjective financial literacy and retail investors' behavior. Journal of Banking and Finance 92 (2018) 168–181 Chhatwani, M., Parija, A. K. (2023). Who invests in cryptocurrency? The role of overconfidence among American investors. Journal of Behavioral and Experimental Economics 107 (2023) 102,107 Clark, R., Lusardi, A., Mitchell, O. (2017). Financial Knowledge and 401(k) Investment Performance. Journal of Pension Economics and Finance 16 (3): 324–47 Fernandes, D., J. G. Lynch Jr, and R. G. Netemeyer (2014). Financial literacy, financial education, and downstream financial behaviors. Management Science 60(8), 1861–1883 Gentile, M., Linciano, N., Lucarelli, C., Soccorso, P. (2015). Financial disclosure, risk perception and investment choice. Evidence from a consumer testing exercise. Consob, Research Department Lusardi, A., Mitchell, O. (2023). The Importance of Financial Literacy: Opening a New Field. Journal of Economic Perspectives 37 (4): 137–54 Van Rooij, M., Lusardi, A., Alessie, R. (2011) Financial literacy and stock market participation. Journal of Financial Economics, 101 (2), 449–472
Retirement planning	Adams, G., Rau, B. (2011). Putting off tomorrow to do what you want today: Planning for retirement. Amer. Psychologist 66(3):180–192 Agnew, J., Bateman, H., Thorp, S. (2013). Financial Literacy and Retirement Planning in Australia. Numeracy 6 (2) Almenberg, J., and Säve-Söderbergh, J. (2011). Financial literacy and retirement planning in Sweden. Journal of Pension Economics and Finance, 10(4), 585–598 Behrman, J. R., Mitchell, O. S., Soo, C. K., Bravo, D. (2012). How financial literacy affects household wealth accumulation. American Economic Review, 102(3), 300–304

(continued)

Table 3.1 (continued)

Financial outcomes	*Sources*
	Bucher-Koenen, T., & Lusardi, A. (2011). Financial literacy and retirement planning in Germany. Journal of Pension Economics & Finance, 10(4), 565–584 Fornero, E., and Monticone, C. (2011). Financial literacy and pension plan participation in Italy. Journal of Pension Economics and Finance, 10(4), 547–564 Gustman, A., Steinmeier, T., Tabatabai, N. (2010). Financial Knowledge and Financial Literacy at the Household Level. National Bureau of Economic Research Working Paper 16,500 Lusardi, A., Mitchell, O. S. (2007). Baby boomer retirement security: The roles of planning, financial literacy, and housing wealth. Journal of Monetary Economics, 54, 205–224 Lusardi, A., Michaud, P., Mitchell, O. (2017). Optimal Financial Knowledge and Wealth Inequality. Journal of Political Economy 125 (2): 431–77 Lusardi, A., Mitchell, O. (2011). Financial Literacy and Planning: Implications for Retirement Well-Being. In Financial Literacy: Implications for Retirement Security and the Financial Marketplace, edited by Olivia S. Mitchell and Annamaria Lusardi, 17–39. Oxford, UK: Oxford University Press Hastings, J., Mitchell, O. (2020). How Financial Literacy and Impatience Shape Retirement Wealth and Investment Behaviors. Journal of Pension Economics and Finance 19 (1): 1–20 Ricci, O., and Caratelli, M. (2017). Financial literacy, trust and retirement planning. Journal of Pension Economics and Finance, 16(1), 43–64 Sekita, S. (2011). Financial literacy and retirement planning in Japan. Journal of Pension Economics and Finance, 10(4), 637–656 Van Rooij, M., Lusardi, A., Alessie, R. (2012). Financial literacy, retirement planning and household wealth. The Economic Journal, 122 (560), 449–478

(continued)

Table 3.1 (continued)

Financial outcomes	*Sources*
Insurance	Allgood, S., Walstad, W. (2016). The effects of perceived and actual financial literacy on financial behaviors. Economic Inquiry, 54, pp. 675–697 Armantier, O., Foncel, J., Treich, N. (2023). Insurance and portfolio decisions: two sides of the same coin? Journal of Financial Economics, 148 (3), pp. 201–219 Bongini, P., Cucinelli, D., Soana, M.G. (2023). Insurance holdings: Does individual insurance literacy matter? Finance Research Letters, Volume 58, Part C, 104,511 Cai, J., De Janvry, A., Sadoulet, E. (2013). Social Networks and the Decision to Insure. Munich Personal RePEc Archive OOWE 46,861 Cole, S., Gine, X., Tobacman, J., Topalova, P., Townsend, R., Vickery, J. (2013). Barriers to Household Risk Management: Evidence from India. American Economic Journal: Applied Economics 5 (1): 104–35 Lin, C., Hsiao, Y. J., Yeh, C. Y. (2017). Financial literacy, financial advisors, and information sources on demand for life insurance. Pacific-Basin Finance Journal, 43, pp. 218–237 Luciano, E., Rossi, M., Sansone, D. (2016). Financial Inclusion and Life Insurance Demand; Evidence from Italian households. CeRP Working Papers from Center for Research on Pensions and Welfare Policies 156/16 Lusardi, A., Mitchell, O. (2014). The economic importance of financial literacy: Theory and evidence. Journal of Economic Literature, 52(1), 5–44 Pitthan, F., De Witte, K. (2021). Puzzles of insurance demand and its biases: A survey on the role of behavioural biases and financial literacy on insurance demand, Journal of Behavioral and Experimental Finance, Volume 30, 100,471 Ricci, O., Santilli, G. (2024). Exploring the link between financial literacy and business interruption insurance: evidence from Italian micro-enterprises. The Geneva Papers on Risk and Insurance - Issues and Practice (2024) 49:663–681 Wang, H., Zhang, D., Guariglia, A., Fan, G. (2021). Growing out of the growing pain': financial literacy and life insurance demand in China. Pacific-Basin Finance Journal, 66, Article 101,459

References

Adams, G., Rau, B. (2011). Putting off tomorrow to do what you want today: Planning for retirement. Amer. Psychologist 66(3):180–192.

Agnew, J., Bateman, H., Thorp, S. (2013). Financial Literacy and Retirement Planning in Australia. Numeracy 6 (2).

Allgood, S., Walstad, W. (2016). The effects of perceived and actual financial literacy on financial behaviors. Economic Inquiry, 54, pp. 675–697

Almenberg, J., and Säve-Söderbergh, J. (2011). Financial literacy and retirement planning in Sweden. Journal of Pension Economics and Finance, 10(4), 585–598.

Armantier, O., Foncel, J., Treich, N. (2023). Insurance and portfolio decisions: two sides of the same coin? Journal of Financial Economics, 148 (3), pp. 201–219.

Atandi, F.G., (2021) "Role of entrepreneur's Competence on Growth of Small and Medium Enterprises", International Journal of Management & Entrepreneurship Research, 3 (2): 84–96.

Babajide, A., Osabuohien, E., Tunji-Olayeni, P., Falola, H., Amodu, L., Olokoyo, F., Adegboye, F., Ehikioya B. (2023) Financial literacy, financial capabilities, and sustainable business model practice among small business owners in Nigeria. Journal of Sustainable Finance & Investment, 13:4, 1670–1692.

Behrman, J. R., Mitchell, O. S., Soo, C. K., Bravo, D. (2012). How financial literacy affects household wealth accumulation. American Economic Review, 102(3), 300–304.

Bellofatto, A., D'Hondt, C., De Winne R. (2018). Subjective financial literacy and retail investors' behavior. Journal of Banking and Finance 92 (2018) 168–181.

Bongini, P., Cucinelli, D., Soana, M.G. (2023). Insurance holdings: Does individual insurance literacy matter? Finance Research Letters, Volume 58, Part C, 104511.

Bucher-Koenen, T., & Lusardi, A. (2011). Financial literacy and retirement planning in Germany. Journal of Pension Economics & Finance, 10(4), 565–584.

Cai, J., De Janvry, A., Sadoulet, E. (2013). Social Networks and the Decision to Insure. Munich Personal RePEc Archive OOWE 46861.

Calcagno, R., Finaldi Russo, P., Galotto, L., Quas, A. (2024). Financial literacy of micro entrepreneurs and access to credit. Bank of Italy, Occasional papers, number 853, June 2024.

Chhatwani, M., Parija, A. K. (2023). Who invests in cryptocurrency? The role of overconfidence among American investors. Journal of Behavioral and Experimental Economics 107 (2023) 102107.

Clark, R., Lusardi, A., Mitchell, O. (2017). Financial Knowledge and 401(k) Investment Performance. Journal of Pension Economics and Finance 16 (3): 324–47.

Clark, R., Lusardi, A., Mitchell, O., (2021). Financial fragility during the COVID-19 pandemic. AEA Papers and Proceedings, 111, 292–296.

Cole, S., Gine, X., Tobacman, J., Topalova, P., Townsend, R., Vickery, J. (2013). Barriers to Household Risk Management: Evidence from India. American Economic Journal: Applied Economics 5 (1): 104–35.

D'Ignazio, A., Finaldi Russo, P., Stacchini, M., (2022). Micro-entrepreneurs' financial and digital competences during the pandemic in Italy, Bank of Italy Occasional Papers No. 724, October 2022.

Dahmen, P. and E. Rodríguez, (2014) "Financial Literacy and the Success of Small Businesses: An Observation from a Small Business Development Center", Numeracy, 7(1).

De Bassa Scheresberg, C. (2013). Financial Literacy and Financial Behavior among Young Adults: Evidence and Implications. Numeracy 6 (2).

Demertzis, M., Domínguez-Jiménez, M., Lusardi, A. (2020). The Financial Fragility of European Households in the Time of COVID-19. Bruegel Policy Contribution 2020/15.

Eniola, A. and H. Entebang, (2017) "SME Managers and Financial Literacy", Global Business Review, 18 (3): 559–576.

Estrada-Mejia C., Diana Mejía, D., Cordoba, P. (2023). Financial literacy and financial wellbeing: Evidence from Peru and Uruguay. Journal of Financial Literacy and Wellbeing (2023), 1, 403–429.

Fernandes, D., J. G. Lynch Jr, and R. G. Netemeyer (2014). Financial literacy, financial education, and downstream financial behaviors. Management Science 60(8), 1861–1883.

Finaldi Russo, P., Galotto L., C. Rampazzi (2022) "The financial literacy of microentrepreneurs: evidence from Italy", Bank of Italy Occasional Papers No. 727, October 2022.

Fornero, E., and Monticone, C. (2011). Financial literacy and pension plan participation in Italy. Journal of Pension Economics and Finance, 10(4), 547–564.

Frisancho, (2023). Spillover Effects of Financial Education: The Impact of School-Based Programs on Parents. Journal of Financial Literacy and Well-being 1: 138–153.

Gentile, M., Linciano, N., Lucarelli, C., Soccorso, P. (2015). Financial disclosure, risk perception and investment choice. Evidence from a consumer testing exercise. Consob, Research Department.

Gustman, A., Steinmeier, T., Tabatabai, N. (2010). Financial Knowledge and Financial Literacy at the Household Level. National Bureau of Economic Research Working Paper 16500.

Hasler, A. Lusardi, A. (2017). The gender gap in financial literacy: A global perspective. Global Financial Literacy Excellence Center, The George Washington University School of Business.
Hastings, J., Mitchell, O. (2020). How Financial Literacy and Impatience Shape Retirement Wealth and Investment Behaviors. Journal of Pension Economics and Finance 19 (1): 1–20.
Hurwitz, A., Mitchell, O., Sade, O. (2021). Longevity Perceptions and Saving Decisions during the COVID-19 Outbreak: An Experimental Investigation. AEA Papers and Proceedings, 111: 297–301.
Hussain, J., Salia, S., Karim, A. (2018). Is knowledge that powerful? Financial literacy and access to finance: an analysis of enterprises in the UK. Journal of Small Business Entrepreneurial Development, 25(6):985–1003.
Klapper, L., Lusardi, A. (2020). Financial literacy and financial resilience: Evidence from around the world. Financial Management 49(3), 589–614.
Kon, Y., Storey, D.J. (2003). A theory of discouraged borrowers. Small Business Economics 21, 37 49.
Lin, C., Hsiao, Y. J., Yeh, C. Y. (2017). Financial literacy, financial advisors, and information sources on demand for life insurance. Pacific-Basin Finance Journal, 43, pp. 218–237.
Luciano, E., Rossi, M., Sansone, D. (2016). Financial Inclusion and Life Insurance Demand; Evidence from Italian households. CeRP Working Papers from Center for Research on Pensions and Welfare Policies 156/16.
Lusardi, A., Michaud, P., Mitchell, O. (2017). Optimal Financial Knowledge and Wealth Inequality. Journal of Political Economy 125 (2): 431–77.
Lusardi, A., Mitchell, O. (2011). Financial Literacy and Planning: Implications for Retirement Well-Being. In Financial Literacy: Implications for Retirement Security and the Financial Marketplace, edited by Olivia S. Mitchell and Annamaria Lusardi, 17–39. Oxford, UK: Oxford University Press.
Lusardi, A., Mitchell, O. (2014). The economic importance of financial literacy: Theory and evidence. Journal of Economic Literature, 52(1), 5–44.
Lusardi, A., Mitchell, O. S. (2007). Baby boomer retirement security: The roles of planning, financial literacy, and housing wealth. Journal of Monetary Economics, 54, 205–224.
Lusardi, A., Mitchell, O. (2023). The Importance of Financial Literacy: Opening a New Field. Journal of Economic Perspectives 37 (4): 137–54.
Pitthan, F., De Witte, K. (2021). Puzzles of insurance demand and its biases: A survey on the role of behavioural biases and financial literacy on insurance demand, Journal of Behavioral and Experimental Finance, Volume 30, 100471.
Ricci, O., and Caratelli, M. (2017). Financial literacy, trust and retirement planning. Journal of Pension Economics and Finance, 16(1), 43–64.

Ricci, O., Santilli, G. (2024). Exploring the link between financial literacy and business interruption insurance: evidence from Italian micro-enterprises. The Geneva Papers on Risk and Insurance - Issues and Practice (2024) 49:663–681.

Roa, M., Garron, I., Barboza, J. (2019). Financial Decisions and Financial Capabilities in the Andean Region. Journal of Consumer Affairs 53: 296–323.

Santos, D., Da Silva, W., Gonzalez, L. (2018). Lower Financial Literacy Induces Use of Informal Loans. RAE– Revista de Administração de Empresas 58: 44–59.

Sekita, S. (2011). Financial literacy and retirement planning in Japan. Journal of Pension Economics and Finance, 10(4), 637–656.

Sotomayor, N., Talledo, J., Wong, S. (2018). Determinants of financial inclusion in Peru: Recent Evidence from the Demand Side. SBS Working Paper No. DT/06/2018.

Van Rooij, M., Lusardi, A., Alessie, R. (2011) Financial literacy and stock market participation. Journal of Financial Economics, 101 (2), 449–472.

Van Rooij, M., Lusardi, A., Alessie, R. (2012). Financial literacy, retirement planning and household wealth. The Economic Journal, 122 (560), 449–478.

Wang, H., Zhang, D., Guariglia, A., Fan, G. (2021). Growing out of the growing pain': financial literacy and life insurance demand in China. Pacific-Basin Finance Journal, 66, Article 101459.

Wise, S. (2013). The Impact of Financial Literacy on New Venture Survival. International Journal of Business and Management; 8 (23): 30–39.

CHAPTER 4

Financial Literacy and Current Challenges: Digitalization and Sustainability

Abstract This final chapter discusses the role of financial literacy in modern challenges such as the adoption of financial services in digital form and the spread of sustainable finance decisions. Specifically, this chapter firstly provides a wide overview of the literature analyzing the relationships between financial literacy, fintech adoption, and sustainable finance, and secondly shows and discusses empirical evidence of these relationships based on research activities conducted in Italy on different datasets and on different targets. The aim of this analysis is to show the positive effects that financial literacy has for the digital transformation and the ESG (Environmental, Social, and Governance) transition of the economy and society, by spreading the adoption of digital technologies in financial services and sustainable investment decisions among individuals and enterprises.

Keywords Financial literacy · Digitalization · Sustainability · Fintech · Sustainable investments

E. Lanciano, *The Relevance of Financial Literacy*,
https://doi.org/10.1007/978-3-031-93397-4_4

4.1 Introduction

In conclusion, the book aims to offer a perspective that also links financial literacy to certain aspects of modernity, such as the digital transformation and the sustainable transition, which, in addition to affecting the economy and society as a whole, have had specific effects in the world of financial services.

Financial services offered in digital form, in fact, are being increasingly used, while for sustainable finance, growing efforts have been seen in recent years, especially from a regulatory perspective, to spread sustainable investment decisions among individuals and businesses. These phenomena have thus assumed an increasing relevance nowadays.

The relationships between financial literacy and the use of digital financial services, and between financial literacy and sustainable finance decisions, are analyzed and discussed in this chapter through empirical evidence, obtained in a national context, but nevertheless found to be robust, significant, and resilient to the use of different measurements and to the analysis of different target populations.

Specifically, the relationship between financial literacy and digital financial services adoption is analyzed in two different samples: a sample of adult people, taken from the survey on financial literacy of Italian households conducted in 2023 by the Italian Financial Education Committee, and a sample of young people, taken from the survey on young people financial literacy conducted in 2023 by Bank of Italy.

The relationship between financial literacy and sustainable finance decisions is also studied among two different samples: the first one is again the sample available from the survey on financial literacy of Italian households conducted by the Italian Financial Education Committee, while the second sample was taken from the survey on financial literacy and digital skills of micro, small and medium entrepreneurs (MSMEs) conducted in 2021 by Bank of Italy, and thus composed of entrepreneurs.

Different measurements of financial literacy are employed, from the Big Three questions and the OECD/INFE toolkit, to measurements developed specifically for the survey goals. In each case, positive findings have emerged, demonstrating how financial literacy can be a useful tool for increasing the adoption of fintech services and spreading sustainable finance decisions.

4.2 Financial Literacy, Digitalization, and Sustainability: A Literature Review

Innovation and sustainability are increasingly important features and crucial modern challenges for the overall economic and social development. In addition, the relevance of these issues has increased especially following the COVID pandemic and the measures introduced by the European National Recovery and Resilience Plan (NRRP) aiming at spreading digital innovation and promoting sustainable development. The segment of the financial services industry has experienced in recent years several transformations related to these trends.

The progressive shift of the financial system from a physical to digital channels makes more important to have adequate financial knowledge and skills, considering the possibilities of making financial decisions independently at any moment and in any place. The rapid technological evolution of the financial sector has caused an increasing range of financial instruments and solutions available and a growing number of financial services providers (Boschmans and Pissareva 2018). In this scenario, financial literacy has thus acquired higher relevance. As mentioned in the second chapter, considering the increasing relevance of the digital innovation in the world of financial services, OECD has recently introduced a measure of digital financial literacy, which combines financial with digital knowledge and skills (Lyons and Kass-Hanna 2021).

Various studies available from the literature provide recent evidence that shows how financial literacy is usually positively associated with digital skills and use of digital financial services. Digital financial literacy is often associated with financial well-being and positive financial behaviors (Rahayu et al. 2022a, 2022b; Sconti 2022; Jhonson et al. 2023), or also with higher investors' awareness of mobile financial services (Morgan and Long 2019; Yoshino et al. 2020; Long et al. 2023). Financial and digital skills have additional relevance for the ability to recognize and avoid digital scams and thus ensure greater financial security (Kass-Hanna et al. 2022; Choung et al. 2023).

Other evidence on this relationship is provided by Ferilli et al. (2024), which found how in the COVID period digital financial literacy and fintech adoption have grown together in EU countries, or by AlSuwaidi and Mertzanis (2024) which found that higher financial literacy significantly drives the fintech market growth across 114 countries in 6 years observed. This relationship does not characterize only individuals but it is confirmed also among entrepreneurs. Nguyen et al. (2024) found in their study that fintech literacy, a very similar concept to digital financial literacy, significantly affect digital entrepreneurial intention through attitudes towards digital entrepreneurship, while D'Ignazio et al. (2022), by analyzing a sample of about 2,000 Italian microentrepreneurs, found that financial literacy, measured using the OECD/INFE toolkit for entrepreneurs, is significantly correlated with the transition to more digitalized business models. These findings are thus consistent among different countries and targets, highlighting the relevant role that financial literacy covers in the transition to a digital financial ecosystem.

The financial system was affected as well by sustainability issues in recent years. Sustainability, in its different meaning, is increasingly affecting the political agenda of countries and international organizations worldwide, and so in the financial industry. The overall development of a sustainable finance ecosystem was indeed supported by a massive regulation process that has produced specific rules aiming at introducing guidance and obligations for companies and financial intermediaries. Mentioning some examples from Europe, the European Commission has launched in March 2018 the Action Plan for Sustainable Finance aiming at promoting investments in sustainable projects and spreading as much as possible the ownership of sustainable financial products such as green, ESG, and sustainability linked loans and bonds (Driessen 2021). This plan includes directives and regulations aiming at defining, for instance, a specific taxonomy for sustainable finance activities or disclosure schemes for companies and financial intermediaries.

Beyond regulation, in order to spread the adoption of these financial decisions among retail investors, it is important also to develop the adequate awareness and knowledge of these financial solutions. For this reason, financial literacy could play a significant role. Very often, indeed, lower levels of financial literacy are associated with wrong conceptions about Socially Responsible Investments (SRI), such as the idea that they underperform the traditional ones, or that they require higher management fees. Therefore, financial literacy is relevant for promoting more informed views about these financial solutions (Meunier and Ohadi 2022).

As well as for the digital financial literacy, the concept of sustainable finance literacy was recently introduced by Filippini et al. (2024). Sustainable finance literacy is defined as the "*knowledge and skill of identifying*

and assessing financial products according to their reported sustainability-related characteristics". The introduction of this concept, which combines financial literacy with sustainable finance elements, shows the growing connection between these two spheres.

The literature analyzing the relationship between financial literacy and sustainable finance or sustainable investments is more divided and thus characterized by conflicting and unconclusive findings. There are, in fact, studies that show a positive effect of financial literacy on sustainable finance decisions, while there are also other studies that, on the contrary, report a negative relationship (Lanciano et al. 2024).

On one side, authors such as Anderson and Robinson (2022) and Gutsche et al. (2021) show respectively how higher financial literacy among Swedish households is positively associated with green financial engagement, knowledge of green investments and environmental values, and how financial literacy is positively related to the individuals' awareness about sustainable investments in Japan. This evidence shows the positive role of financial literacy and education also for developing awareness about sustainability and promoting sustainable investments (Lee-Ying et al.2022).

On the other side, examples of negative relationship between financial literacy and sustainable finance are reported by D'Hondt et al. (2022), Riedl and Smeets (2017), or Rossi et al. (2019), which found that negative associations emerge between financial literacy and the exposure of the stock portfolio to sustainability factors, that more financially literate people own less sustainable finance products, and that financially literate individuals are less interested in SRI.

Thus, regarding the effects of financial literacy on sustainable finance decisions, surely further evidence is needed in order to better understand this relationship, considering the divergent findings reported in literature in various contexts.

4.3 Financial Literacy and Fintech Adoption

This section aims to provide further empirical evidence on the relationship between financial literacy and the use of financial services in digital form. These findings are based on research activities conducted on Italian samples. The goal is to show that this relationship is persistent among multiple databases and using different measurements of financial literacy and fintech adoption. The data used for this analysis were taken from two different samples.

The first sample analyzed is available from the fourth annual survey on the level of financial literacy of Italian households conducted by the Italian Financial Education Committee (Edufin Committee) in 2023. It is composed of 5,002 respondents from Italy aged more than 18 years old, thus being quite large and nationally representative. Additionally, the sample contains several sociodemographic information of respondents such as their gender, age, geographical origin, educational level, and income.

In order to analyze the relationship between financial literacy and the use of digital financial services, a measure of observed financial literacy is employed, given by the number of correct answers to the Big Three financial literacy questions, regarding the topics of interest rate, inflation, and risk diversification. Each of these questions assumes value 1 in case of correct answer while 0 otherwise. Thus, the score of financial literacy can assume values from 0 to 3. In this sample, only 44.5% of the respondents answer correctly to all of these three questions.

Regarding the adoption of fintech services, in this survey, it is possible to assess the individual confidence in adopting fintech services through a set of questions asking to the respondents how much they are comfortable on a 1–4 scale in using digital devices for making payments, transferring money, managing a bank account, obtaining a loan, signing an insurance contract, investing in pension funds, protecting data when using financial services and finally purchasing financial products. The answers to these questions assume a positive value when the respondents declare to feel high or medium–high confidence, while otherwise it is 0. The overall score of fintech adoption thus ranges between 0 and 8. The questions aimed at investigating fintech adoption are reported in the following table (Table 4.1).

Finally, sociodemographic control variables, typically employed in studies dealing with financial literacy, are included in the investigation such as gender, geographical area, age, education, and income of the respondents. Precisely, three geographical areas (north, center, south and islands), age groups (young 18–44, adult 45–64, elderly 65+), three educational levels (university, high school, lower education), and three monthly income levels (more than € 2,455, from € 1,265 to € 2,454, less than € 1,265) are considered.

Table 4.1 Fintech adoption questions

How comfortable are you with using your smartphone, PC, or other digital devices to...

		High	*Medium–high*	*Medium–low*	*Low*
a.	Transfer money/make payments	1	2	3	4
b.	Manage a bank account	1	2	3	4
c.	Find out how to get a loan	1	2	3	4
d.	Sign an insurance contract	1	2	3	4
e.	Join supplementary pension plans	1	2	3	4
f.	Pay with a card	1	2	3	4
g.	Protect sensitive data when purchasing financial services or make payments online	1	2	3	4
h.	Purchase financial instruments through online trading platforms or robo advisors	1	2	3	4

Considering the goal of this research activity, an Ordinary Least Square (OLS) regression model is employed, with Fintech adoption as dependent variable, and Financial literacy as main independent variable of interest. In addition, sociodemographic control variables are included in the model, as indicated by $\sum \delta_i$. The model is represented below (Model 1):

$$(1)\text{Fintech adoption}_i = \beta_0 + \beta_1 \text{Financial Literacy}_i + \sum \delta_i + \varepsilon_i$$

Results from this model are represented in the following table (Table 4.2), which reports coefficients indicating the nature of the relationships between the variables included in the model. These results show how financial literacy has a positive and highly significant effect ($p < 0.01$) on the adoption of financial services in digital form. It is also interesting that among the different sociodemographic categories, males, young, and high-income respondents are more likely to adopt fintech services. These differences are in line with the typical sociodemographic differences

Table 4.2 Financial literacy and fintech adoption in adult people

	(1)
Variables	Fintech adoption
Financial literacy	0.251***
	(0.0360)
Male	0.430***
	(0.0651)
North	−0.0640
	(0.0709)
Centre	0.00140
	(0.0857)
Young	1.149***
	(0.110)
Adult	0.697***
	(0.105)
University	0.156
	(0.115)
High school	0.115
	(0.113)
High income	1.019***
	(0.0908)
Medium income	0.668***
	(0.0859)
Constant	3.472***
	(0.165)
Observations	5,002
R-squared	0.085

Robust standard errors in parentheses
***$p < 0.01$, **$p < 0.05$, *$p < 0.1$

recorded in other studies or global investigations dealing with financial literacy and previously discussed.

The second sample investigated in this section was selected from the 2023 Bank of Italy survey on young people's financial literacy,[1] and is composed by 5372 survey respondents, aged between 18 and 34, nationally representative of the Italian young people population. Considering the same goals, this second analysis is set and conducted in a very similar way to the previous one. In this case, the employed measurement of financial literacy is quite different. This proxy is composed of the amount of

[1] https://www.bancaditalia.it/pubblicazioni/indagini-alfabetizzazione/2023-indag-alfabetizzazione-giovani/index.html.

correct answers to a set of 16 financial literacy survey questions, 8 financial knowledge and 8 financial behavior questions, including the Big Three topics (interest rate, inflation and risk diversification) and other financial topics such as risk-return relationship, mortgages, savings, retirement planning, and budgeting. The questions used for this measurement of financial literacy are reported in the following table (Table 4.3).

For measuring the adoption of fintech services, this survey employs the same questions used in the previous investigation, thus asking to the respondents how much they feel comfortable in using financial services (payments, bank account, credit, insurance, trading, advisory) in a digital form. Finally, the analysis includes a set of sociodemographic control variables such as gender, age (three age groups), geographical area (three geographical areas), municipal dimension (three dimensions), education (three educational levels), wealth (three wealth levels), and profession (three profession groups). The methodology is the same of the previous analysis, thus based on an OLS model with the adoption of fintech services considered as dependent variable, while financial literacy and sociodemographic controls as independent variables. Results of this second investigation are reported in the following table (Table 4.4).

Financial literacy has again a positive and highly significant ($p < 0.01$) effect on the adoption of financial services in digital form. In this case too, there are some sociodemographic differences: fintech services are more likely to adopted by male, worker, highly educated, wealthy respondents, and individuals belonging to the oldest age group of this survey (30–34).

These results show that individuals with higher financial literacy are more likely to use essential financial services, such as bank accounts, payments, insurance, or investments, in a digital form. Thus, considering the overall evidence emerged in this section, it seems that generally financial literacy can play a relevant role in developing the right awareness and spreading the use of digital financial services. These results are in line with the previous literature on this relationship and, considering also the findings on entrepreneurs (D'Ignazio et al. 2022), they are persistent among multiple investigations, different samples, and various measurements of financial literacy.

Table 4.3 Financial literacy questions (Bank of Italy survey on young people's financial literacy)

Topic	*Question*	*Answer*
Financial knowledge		
Interest rate	Suppose to have €100 on your banking account, with a 2% annual interest rate. How much will you have in your account after 1 year?	€102
Inflation	Five brothers receive a total of EUR 1,000 as a gift today but have to wait a year to dispose of their share. If the annual inflation rate is 8%, in a year's time each brother will be able to buy with their own sum:	Less than today
Risk diversification	It is usually possible to reduce investment risk by purchasing securities and shares of many types	True
Mortgages	A 15-year mortgage usually requires the payment of higher installments than a 30-year mortgage, but the total interest paid during the total duration of the mortgage is lower	True
Compound interest	Suppose to have €100 on your banking account, with a 2% annual interest rate. How much will you have in your account in 5 years?	More than €110
Risk-return relationship	An investment with a high return is probably low risk	False
Stock market predictability	A financial advisor can predict with extreme accuracy the performance of a stock share	Disagree/ Strongly disagree
Inflation protection	By keeping my savings in my current account, I am always protected from the risk that inflation will reduce its value	Disagree/ Strongly disagree
Financial behavior		
Savings	Do you normally manage to save something at the end of the month?	Always/ Often
Supplementary pension	How do you plan to cope with the needs you will have when you retire? With a supplementary pension fund	Yes
Retirement planning	When do you think it would be better to start saving for retirement?	As soon as possible, even small amounts
Budgeting	Before buying something, I consider whether I can afford it	Always/ Often
Budgeting	How often do the following situations happen to you? Realizing that you have spent more than you had planned	Never

(continued)

Table 4.3 (continued)

Topic	*Question*	*Answer*
Respect of deadlines	How often do the following situations happen to you? Realizing that you have forgotten a payment deadline	Never
House expenses	Could you quantify how much you spend each month in total on the services you have on your subscription (e.g. transport, TV streaming services, music, gaming, telephony, etc.)?	Yes, precisely
Long-term goals	I set myself long-term financial goals and I commit myself to achieving them	Agree/ Strongly agree

4.4 Financial Literacy and Sustainable Finance Decisions

This final section aims to provide empirical evidence on the role that financial literacy may cover in fostering sustainable finance decisions. In this case too, these research activities are conducted in Italy on two different samples.

In the first investigation, this relationship is analyzed considering the sample available from the 2023 survey on financial literacy of Italian households conducted by the Italian Financial Education Committee, composed of 5,002 observations, previously used for studying the relationship between financial literacy and fintech adoption. This survey, indeed, contains also specific questions analyzing the level of knowledge of sustainable finance concepts such as the definition of sustainable finance, the ESG factors, and the definition of sustainable development, aiming at assessing a sort of sustainable finance literacy of the respondents. These three questions precisely ask to the respondents to declare how much they know about these three concepts on a 1–4 scale, where 1 is "never heard", 2 is "just heard", 3 is "basic knowledge", and 4 is "advanced knowledge". Thus, these questions are all considered as categorical variables with values from 1 to 4, and are set as dependent variables of this analysis (sustainable finance, ESG, sustainable development).

Analyzing the answers to these questions provided by the respondents, an overall limited knowledge of these topics among Italian households emerges but distributed differently among the questions. For all these three concepts, only a few hundred of the respondents claim to

Table 4.4 Financial literacy and fintech adoption in young people

	(1)
Variables	Fintech adoption
Financial literacy	0.189***
	(0.00918)
Male	0.215***
	(0.0511)
University	0.173*
	(0.101)
High school	0.229**
	(0.0934)
Young (18–23)	−0.362***
	(0.0738)
Medium age (24–29)	−0.130**
	(0.0601)
North	−0.0855
	(0.0584)
Center	−0.0599
	(0.0695)
Student	0.144*
	(0.0738)
Worker	0.565***
	(0.0675)
High wealth	0.435***
	(0.0832)
Medium wealth	0.279***
	(0.0669)
Constant	2.195***
	(0.127)
Observations	5,372
R-squared	0.152

Robust standard errors in parentheses
***$p < 0.01$, **$p < 0.05$, *$p < 0.1$

have advanced knowledge, while analyzing the questions separately, ESG factors is the least known topic, while the one with the most positive responses is the definition of sustainable development. At an intermediate level is sustainable finance. These findings are reported in the following table (Table 4.5).

Financial literacy, instead, is measured again with the Big Three questions and is set as main independent variable of interest. Finally, the same set of sociodemographic control variables ($\sum \delta_i$) is included (gender,

Table 4.5 Knowledge of sustainable finance concepts

	Never heard	*Just heard*	*Basic knowledge*	*Advanced knowledge*
Sustainable finance	1,082	2,051	1,546	323
ESG	2,122	1,581	1,037	262
Sustainable development	422	1,947	2,017	616

Source Author's elaboration of data from the 2023 survey on financial literacy of Italian households

geographical area, age, education and income of the respondents). The relationship between these variables is assessed through an OLS regression model, expressed as follows (Model 2).

$$(2)\text{Sustainable Finance}_i = \beta_0 + \beta_1 \text{Financial Literacy}_i + \sum \delta_i + \varepsilon_i$$

Sustainable finance indicates each time a different dependent variable (Sustainable finance, ESG, Sustainable development). Results of this model are reported in Table 4.6. The coefficients and the p-values show a positive and highly significant effect of financial literacy in knowledge of topics related to sustainable finance. The relationship is persistent among all the three concepts. Regarding control variables, the role of education and wealth immediately emerges. In fact, respondents with higher education and income are significantly more likely to know these topics. Finally, for two of the three concepts, higher knowledge emerges among male respondents.

The second investigation that analyzes the relationship between financial literacy an sustainable finance is conducted among entrepreneurs. Precisely, a sample of micro, small, and medium entrepreneurs (MSMEs) is analyzed. This sample is available from the 2021 Bank of Italy survey on financial competences of entrepreneurs,[2] which is part of a wider OECD/INFE project aiming at gathering international data on the financial literacy from firms of small dimensions. This sample is composed of 1,998 Italian non-financial firms with less than 50 employees. The survey includes also information on business characteristics such as size, turnover,

[2] https://www.bancaditalia.it/statistiche/tematiche/indagini-famiglie-imprese/alfabetizzazione-imprese/index.html?com.dotmarketing.htmlpage.language=1.

Table 4.6 Financial literacy and knowledge of sustainable finance concepts

	(1)	*(2)*	*(3)*
Variables	Sustainable finance	ESG	Sustainable development
Financial literacy	0.100***	0.0340**	0.165***
	(0.0130)	(0.0135)	(0.0125)
Male	0.159***	0.123***	0.00514
	(0.0249)	(0.0268)	(0.0236)
North	0.0529*	0.0172	−0.00428
	(0.0272)	(0.0292)	(0.0259)
Center	−0.0103	−0.0516	−0.00671
	(0.0332)	(0.0351)	(0.0320)
Young	−0.0552	0.0216	−0.0183
	(0.0394)	(0.0412)	(0.0374)
Adult	−0.0423	−0.00580	−0.00773
	(0.0362)	(0.0380)	(0.0343)
University	0.376***	0.389***	0.301***
	(0.0435)	(0.0443)	(0.0413)
High school	0.143***	0.132***	0.0862**
	(0.0416)	(0.0420)	(0.0396)
High income	0.192***	0.140***	0.0909***
	(0.0339)	(0.0358)	(0.0323)
Medium income	0.104***	0.0460	0.0362
	(0.0316)	(0.0325)	(0.0305)
Constant	1.585***	1.436***	2.000***
	(0.0596)	(0.0596)	(0.0579)
Observations	5,002	5,002	5,002
R-squared	0.073	0.042	0.074

Robust standard errors in parentheses
***$p < 0.01$, **$p < 0.05$, *$p < 0.1$

industry, geographical location, and sociodemographic characteristics of the entrepreneurs such as gender, age, and level of education.

In this case, being an OECD/INFE project, financial literacy is measured with the OECD/INFE toolkit for the entrepreneurs (OECD 2020). Thus, the questionnaire includes a set of questions aiming at assessing the individual level of financial literacy of the entrepreneurs composed of three sections: five questions on financial knowledge, nine questions on financial behavior, and three questions on financial attitude. Thus, the overall score of financial literacy is obtained by the sum of the correct answers to these questions, and ranges between 0 and 17. The questions cover several topics: knowledge of dividends, equity, mortgages, inflation, risk-return relationship, management of business bank account, business accounting, strategies to cope with theft, business data protection, business long-term planning, business financing, and profitability.

Italian entrepreneurs show an average financial literacy score of 2.15 out of 3, which corresponds to 71.6% of correct answers. Considering that OECD sets the 80% of correct answers as a sufficient score for being considered as financially literate,[3] the overall level of financial literacy of this sample of Italian entrepreneurs can be considered as quite limited. Higher percentages of correct answers are reported for financial behavior (78%) and knowledge (72%), while this percentage decreases for financial attitude (65%). This evidence on financial literacy is reported in Table 4.7, which shows summary statistics of the overall financial literacy score and its components.

This survey contains also specific questions about sustainable finance decisions related to the business. For this study, two specific questions were selected, that ask to the respondents whether they take into account environmental and social impacts when they make investments for the business. Therefore, in this case, impact investment decisions are considered as sustainable finance variables. These two questions assume value 1 when the respondents declare to be agree or strongly agree with these statements, while otherwise they are equal to 0. They are thus considered

Table 4.7 Financial literacy summary statistics

Variable	*N*	*Mean*	*SD*	*Min*	*p25*	*Median*	*p75*	*Max*
Financial literacy	1,998	2.15	0.57	0.00	1.82	2.24	2.56	3.00
Financial knowledge	1,998	0.72	0.27	0.00	0.60	0.80	1.00	1.00
Financial behavior	1,998	0.78	0.19	0.00	0.67	0.78	0.89	1.00
Financial attitude	1,998	0.65	0.31	0.00	0.33	0.67	1.00	1.00

Source Author's elaboration of data from the 2021 Bank of Italy survey on financial competences of entrepreneurs

[3] For more details please consult the OECD/INFE survey instrument to measure the financial literacy of MSMEs (OECD 2020).

Table 4.8 Impact investment questions

When I make an investment for the business, I take into account its environmental impact	*Total sample*	*Percent*
0 Otherwise	524	26.23
1 Agree/strongly agree	1,474	73.77
Total	1,998	100.00
When I make an investment for the business, I take into account its social impact		
0 Otherwise	600	30.03
1 Agree/strongly agree	1,398	69.97
Total	1,998	100.00

Source Author's elaboration

as variables of binary nature and set as dependent variables. The wording of these questions together with the details of the answers is reported in the following table (Table 4.8).

In order to analyze the relationship between the financial literacy score and sustainable finance decisions, represented by investments with environmental and social impacts, the same OLS model of the previous analysis has been employed, where this time sustainable finance represents first investments with environmental impacts (Environmental impact) and then investments with social impacts (Social impact). Finally, a set of control variables is included, related to business characteristics (business turnover, industry, geographical location) and sociodemographic information of the entrepreneur (gender, age, education). Results of this model are reported in the following table (Table 4.9). The coefficients show that financial literacy has a strong effect on both business investment decisions with environmental and social impact. There are no particularly relevant findings among the control variables, with only some differences between industries and geographical locations.

Thus, this evidence shows how financial literacy can positively affect sustainable finance decisions among entrepreneurs too.

Summarizing the findings emerging from this section, the two research activities show first that financial literacy may positively affect the degree of knowledge of sustainable finance, and then that financial literacy may affect also sustainable investments, thus promoting the spread of these decisions among retail investors and contributing to the sustainable transition of the economy and the society. These relationships may have

Table 4.9 Financial literacy and impact investment decisions

	(1)	*(2)*
Variables	Environmental impact	Social impact
Financial literacy	0.0460***	0.0517***
	(0.00328)	(0.00326)
Turnover	−0.00932	−0.00968
	(0.00755)	(0.00799)
Agriculture	0.142***	0.0294
	(0.0433)	(0.0474)
Manufacturing	0.0391	0.00959
	(0.0483)	(0.0490)
Real estate	−0.0772	−0.0393
	(0.0486)	(0.0476)
Retail and wholesale trade	−0.00792	−0.0156
	(0.0388)	(0.0389)
Transportation	0.0445	0.0329
	(0.0603)	(0.0628)
Accommodation	0.0577	−0.0283
	(0.0474)	(0.0527)
Personal services	0.0191	−0.0179
	(0.0418)	(0.0419)
Information and communication	−0.0375	−0.151**
	(0.0661)	(0.0748)
Business services	−0.0405	−0.0273
	(0.0384)	(0.0382)
Male	0.0119	0.0177
	(0.0211)	(0.0218)
North	−0.0301	−0.0714***
	(0.0219)	(0.0226)
Centre	−0.0627**	−0.0521*
	(0.0268)	(0.0274)
Age	−0.00006	−0.00117*
	(0.000670)	(0.000692)
University	0.0276	0.0360
	(0.0325)	(0.0345)
High school	0.00270	0.0141
	(0.0296)	(0.0309)
Constant	0.206***	0.174**
	(0.0673)	(0.0680)
Observations	1,998	1,998
R-squared	0.105	0.114

Robust standard errors in parentheses
***$p < 0.01$, **$p < 0.05$, *$p < 0.1$

different explanations. It would be difficult to say that financial literacy leads to greater public and social interest or to a better understanding and knowledge of issues such as the global climate change. The connection between financial literacy and sustainability may lie in the fact that financial literacy increases the awareness of financial products and thus may lead individuals to make more sophisticated financial decisions, such as sustainable investments, or considering also that financial literacy and sustainability may have some behavioral attitudes in common, such as a sense of responsibility and awareness towards management of financial resources for financial literacy and in considering the impacts of decisions for sustainability.

The goal of these investigations was indeed to show that this relationship has the same positive nature among different samples and employing two different measurements of financial literacy (Big Three and OECD/INFE toolkit). These multiple estimations, as well as in the section where the relationship between financial literacy and fintech adoption is analyzed, provide robustness to these results.

While for financial literacy and fintech adoption, these results are in line with the overall evidence from the existing literature on these topics, for the relationship between financial literacy and sustainable finance certainly further investigations are needed considering the conflicting opinions emerging from the literature.

4.5 Conclusions

This final chapter sets multiple goals for the reader. The first is to show how financial literacy can become relevant not only for the financial decisions and outcomes seen in the previous chapter but also for important challenges nowadays such as digitization and sustainability.

Indeed, the results show the potential positive role of financial literacy for the development of these phenomena, clearly limited to the area of knowledge of financial products, use of financial services, and investment decisions.

The second goal is to provide the reader with an example of applied research in the field of financial literacy. In fact, financial literacy studies are typically set up in a similar way, with the effect of financial literacy on a given financial decision or outcome being estimated through a regression model. In addition, the control variables used are generally those that are included in these types of studies and represent the factors that typically can affect the individual level of financial literacy. Finally, these research activities represent also an example of employment of different measurements of financial literacy.

Digitalization and sustainability are hot topics of the moment in the field of financial literacy research. This is also evident from the significant production of literature in recent years that links financial literacy to these topics, as revealed by the literature review contained in this chapter. The positive link between financial literacy and these aspects emphasizes especially, beyond the importance for the well-being of the individual, the relevance for the overall society, as they represent contemporary challenges to which all social actors, from individual citizens to public and governmental bodies, are called upon to contribute.

References

AlSuwaidi, R., Mertzanis, C. (2024). Financial literacy and FinTech market growth around the world, International Review of Financial Analysis, Volume 95, Part B, 103481.

Anderson, A., Robinson, D. T. (2022). Financial Literacy in the Age of Green Investment. Review of Finance, 1551–1584.

Boschmans, K., L. Pissareva, (2018). Fostering Markets for SME Finance: Matching Business and Investor Needs, OECD SME and Entrepreneurship Papers, No. 6, OECD Publishing, Paris.

Choung, Y., Chatterjee, S., Pak, T.Y. (2023). Digital financial literacy and financial well-being. Finance Research Letters, 58(B):104438.

D'Hondt, C., Merli, M., & Roger, T. (2022). What drives retail portfolio exposure to ESG factors? Finance Research Letters, 46.

D'Ignazio, A., Finaldi Russo, P., Stacchini, M., (2022). Micro-entrepreneurs' financial and digital competences during the pandemic in Italy, Bank of Italy Occasional Papers No. 724, October 2022.

Driessen, M. (2021). Sustainable Finance: An Overview of ESG in the Financial Markets. In: Busch, D., Ferrarini, G., Grünewald, S. (eds) Sustainable Finance in Europe, 320-350. EBI Studies in Banking and Capital Markets Law. Palgrave Macmillan, Cham.

Ferilli, F., Palmieri, E., Miani, S., Stefanelli, V. (2024). The impact of FinTech innovation on digital financial literacy in Europe: Insights from the banking industry, Research in International Business and Finance, Volume 69, 102218.

Filippini, M., Leippold, M., Wekhof, T. (2024). Sustainable Finance Literacy and the Determinants of Sustainable Investing. Journal of Banking and Finance, 163, 107167.

Gutsche, G., Nakai, M., & Arimura, T. H. (2021). Revisiting the determinants of individual sustainable investment: the case of Japan. Journal of Behavioral and Experimental Finance, 30.

Jhonson, B., Andriani, R., Noviana, I., Tamara, D. (2023) The influence of digital financial literacy on financial well-being through spending, saving, and investment behavior in Indonesia. Journal of Business Studies and Management Review, 6(2):157–168.

Kass-Hanna, J., Lyons, A.C., Liu, F. (2022). Building financial resilience through financial and digital literacy in South Asia and Sub-Saharan Africa. Emerging Markets Review, 51:100846.

Lanciano, E., Previati, D., Ricci, O., Santilli, G. (2024). Financial literacy and sustainable finance decisions among Italian households, Journal of Economics and Business, 106220.

Lee-Ying, T., Hen-Toong, T., Gek-Siang. T. (2022). Digital financial inclusion: A gateway to sustainable development. Heliyon, 8, e09766.

Long, T.Q., Morgan, P.J., Yoshino, N. (2023). Financial literacy, behavioral traits, and ePayment adoption and usage in Japan. Financial Innovation, 9(1):1–30.

Lyons, A.C., Kass-Hanna, J. (2021) A methodological overview to defining and measuring "digital" financial literacy. Financial Planning Review, 4(2), e1113.

Meunier, L., & Ohadi, S. (2022). Misconceptions about socially responsible investments. Journal of Cleaner Production, 373, 133868.

Morgan, P.J., Long, T.Q. (2019). Fintech and financial literacy in The Lao PDR. Asian Development Bank Institute. ADBI Working Paper no. 933.

Nguyen, T., Dao, T., Tran, T., Nguyen, H., Ngoc Le, L., Pham, N. (2024). Fintech literacy and digital entrepreneurial intention: Mediator and Moderator Effect, International Journal of Information Management Data Insights, Volume 4, Issue 1, 100222.

OECD (2020). OECD/INFE survey instrument to measure the financial literacy of MSMEs.

Rahayu, R., Ali, S., Aulia, A., Hidayah, R. (2022a). The current digital financial literacy and financial behavior in Indonesian millennial generation. Journal of Accounting and Investment 23(1):78–94.

Rahayu, R., Juita, V., Rahman, A., Fitriamiranti, S., Rafles, R. (2022b). The level of digital financial literacy and financial well-being of people in West Sumatra Indonesia. Operations Management and Information System Studies 2(2):66–76.

Riedl, A., Smeets, P., (2017). Why do investors hold socially responsible mutual funds? Journal of Finance 72 (6), 2505–2550.

Rossi, M., Sansone, D., Van Soest, A., Torricelli, C. (2019). Household preferences for socially responsible investments. Journal of Banking and Finance, 105, 107–120.

Sconti, A. (2022). Having Trouble Making Ends Meet? Financial Literacy Makes the Difference. Italian Economic Journal:1–32.

Yoshino, N., Morgan, P.J., Long, T.Q. (2020). Financial literacy and fintech adoption in Japan. Asian Development Bank Institute. ADBI Working Paper no. 1095. Tokyo, Asian Development Bank Institute.

Final Remarks

The main purpose of this book is to provide a general knowledge and awareness about the topic of financial literacy. Additionally, the book tries to explain why financial literacy should be relevant for the individual and social well-being. Very often, indeed, people know what financial literacy is or know the basic financial literacy concepts, but they are not totally aware of the relevance of this topic, because they are not informed of the consequences of being illiterate, or they are not aware of global and sociodemographic differences in the level of financial literacy that may cause geographical or social inequalities among the most vulnerable groups.

In order to provide this general overview of the topic, the book begins with showing and explaining the main definitions of financial literacy and education, and the related topics to know for being considered as financially literate such as, for instance, interest rate, inflation, risk diversification, risk-return relationship, mortgages, insurance, or pensions. The definitions' explanations allow to understand that these are quite complex topics, considering that they are not only a matter of knowledge but include several other aspects such as behavioral, decisional, and attitudinal aspects. Following this first part of the chapter, global evidence of the level of financial literacy taken from international surveys are provided, together with a deeper analysis of the differences between sociodemographic categories, which shows how financial literacy usually reflects the level of the overall economic development of countries and how generally

E. Lanciano, *The Relevance of Financial Literacy*,
https://doi.org/10.1007/978-3-031-93397-4

financial literacy is lower among women, young, elderly, less educated and less wealthy people. Finally, the chapter explains which are the key policy actors for financial literacy strategies.

The following chapter is more focused on technical aspects, by providing an overview of the main financial literacy measurements employed in literature and in global surveys, such as the Big Three or Five questions or the OECD/INFE toolkit. This chapter also provides an explanation of typical limits and technical problems related to these measurements and finally includes alternative views and research approaches developed over the years.

The book proceeds with an analysis of the relationships between financial literacy and related financial outcomes, such as savings and financial resilience, access to credit, investment decisions, retirement planning, and insurance, aiming at showing the relevance of financial literacy for these decisions. The analysis of these relationships was conducted through a wide review of some of the most notable studies available in the literature on these topics. This work may also be useful for defining a set of studies associated with specific relationships between financial literacy and financial outcomes.

In the final chapter, recent research evidence on the role of financial literacy for modern challenges such as digitalization and sustainability are provided, which shows how financial literacy could be a relevant tool also for the digital and sustainable transition of the economy and society. These findings show the positive effect that financial literacy has on the adoption of fintech services and in promoting sustainable finance decision. Even if at country level, as financial literacy studies are usually conducted, these findings are robust to multiple measurements, different samples and are highly statistically significant.

The book therefore includes various theoretical and technical aspects aiming at developing an overall knowledge and awareness of the topic, by, at the same time, offering various international evidence, discussing the main research approaches, reviewing the literature, and finally providing a research focus on contemporary and emerging issues, and thus showing why financial literacy should be considered as increasingly relevant for empowering individuals and society. The book can be useful for various ranges of readers, from individuals who want to make a first approach to the topic, to people who already know the topic and want to learn something more on research aspects or want to know an overview of the available literature on the topic and on specific relationships.

The book also emphasizes the importance of being aware of the relevance of financial literacy, because it is not enough to know the topic or the related concepts, but it is important to develop awareness of the effects that financial literacy has on the financial decisions of individuals or entrepreneurs. For this last aspect, the role of the key actors who are committed in the development of financial literacy assumes a crucial importance, especially in terms of setting policies and organizing initiatives with the aim of raising this awareness of the topic, possibly starting from a young audience, so that this awareness can be developed already at an early age. Improving financial literacy from a young age, indeed, will not only bring short-term benefits to individual money management but will also help people to make smarter financial decisions in the long-term, in order also to prevent the development of future social inequalities.

Index

A
Age, 11, 13, 17
Alternative, 26, 30, 32–34
Approaches, 24, 32, 33, 35, 36
Attitude, 3, 4, 8–10
Authorities, 2, 15–18
Awareness, 43, 45, 47

B
Behavior, 2, 8–10
Big Five, 24–27, 32, 33, 36
Big Three, 24–27, 32, 33, 36

C
Control variables, 68, 69, 71, 74, 75, 78, 80
Countries, 2, 5–16, 18
Credit, 42–46, 54

D
Data, 68, 69, 75, 77
Decisions, 3, 4, 8, 11, 14, 18, 19
Digitalization, 65
Disciplines, 24, 35

E
Education, 2, 4, 11, 14, 15, 17, 18
Endogeneity, 24, 30, 31
Entrepreneurs, 24, 27–29
Environmental impact, 78, 79
Evidence, 49–51, 53
Experimental, 24, 34, 35

F
Financial decisions, 42, 43, 47, 48, 51
Financial education, 3–5, 10, 13, 16–18
Financial literacy, 2–18, 23, 24, 26–36, 42–52, 57, 64–81
Financial outcomes, 49, 51–57
Financial services, 64, 65, 67–69, 71, 80

E. Lanciano, *The Relevance of Financial Literacy*,
https://doi.org/10.1007/978-3-031-93397-4

Fintech, 64–66, 68–71, 73, 74, 80

G
Gender, 11, 12
Geographical areas, 5–7, 11
Governments, 2, 15, 16, 18

I
Income, 11, 14
Individuals, 2–8, 11, 15, 17, 18, 24, 26–31, 34, 35
INFE, 27, 28, 30, 33, 36
Inflation, 4, 5, 7, 8
Instrumental variable (IV), 24, 30–32
Insurance, 42, 43, 49, 50, 57
Interest rate, 4, 5, 7, 8
International, 2, 5, 15
Investment, 42, 43, 46–48

K
knowledge

L
Literature, 42, 43, 47, 49, 51, 65, 67, 71, 80, 81

M
Measurement, 24, 27–29, 36
Measurement error, 24, 30

N
National strategies, 15–18

O
OECD, 2, 3, 5, 8–11, 13–18, 27–30, 32, 33, 36
Overconfidence, 47, 55

P
Perspectives, 24, 33–35
Planning, 43, 45, 48, 49, 53, 55
Policymakers, 15

Q
Questions, 24–30, 32–34, 36

R
Regression, 69, 75, 80
Relevant, 44, 45, 47, 50, 51
Research, 23, 24, 32–36, 67, 69, 73, 78, 80, 81
Retirement, 42, 43, 48, 49, 53, 55, 56
Risk-diversification, 4, 5, 7, 8
Risk propensity, 47

S
Saving, 42–44, 48, 49, 52
Score, 6–12, 14
Skills, 3–5, 8, 10, 13, 14
Social impact, 77–79
Sociodemographic, 68–71, 74, 76, 78
Sociodemographic differences, 8, 11
Survey, 2, 5, 7–11, 14, 16–18, 64, 68, 70–73, 75, 77
Sustainability, 65–67, 80, 81
Sustainable finance, 64, 66, 67, 73–78, 80
Sustainable investments, 64, 67, 78, 80

T
Toolkit, 27–30, 33, 36

U
Understanding, 2–8, 10, 13

V
Variables, 69–71, 73–77, 79

W
Well-being, 2–4

Y
Young, 2, 13, 14, 17, 18

GPSR Compliance

The European Union's (EU) General Product Safety Regulation (GPSR) is a set of rules that requires consumer products to be safe and our obligations to ensure this.

If you have any concerns about our products, you can contact us on ProductSafety@springernature.com

In case Publisher is established outside the EU, the EU authorized representative is:

Springer Nature Customer Service Center GmbH
Europaplatz 3
69115 Heidelberg, Germany

Batch number: 08713873

Printed by Printforce, the Netherlands